SpringerBriefs in Applied Sciences and Technology

More information about this series at
http://www.springer.com/series/8884

Ron Hanifan

Perfecting Engineering and Technical Drawing

Reducing Errors and Misinterpretations

 Springer

Ron Hanifan
Del Rio, TN
USA

Portions of this Work previously appeared in Reduce Your Engineering Drawing Errors : Preventing the Most Common Mistakes, 978-1-606-50210-5 (originally published in 2010 by Momentum Press).

ISSN 2191-530X   ISSN 2191-5318 (electronic)
ISBN 978-3-319-06982-1 ISBN 978-3-319-06983-8 (eBook)
DOI 10.1007/978-3-319-06983-8
Springer Cham Heidelberg New York Dordrecht London

Library of Congress Control Number: 2014945752

Printed on acid-free paper

Springer is part of Springer Science+Business Media (www.springer.com)

This book is dedicated to my lovely wife, Catherine. She is the inspiration and driving force behind all that I write. She gave me the encouragement to finish this book and to leave my little mark in this beautiful world. Semper Fi.
Ron Hanifan

Preface

After checking thousands of drawings, it is evident that not much progress has been accomplished in complying with proper documentation of drawings and drawing selection. Most customers and/or design activities are not familiar enough with the established rules, even though they haven't changed in many years. The same geometric dimension and tolerancing errors are present today as they were 60 years ago. The intent of dimensioning and tolerancing has not really changed over the years except that definitions have become much more exact and many more seldom used applications have been added.

Most companies that I have been associated with justify their noncompliance by using the justification of "That is the way we always have done it," "We have done it that way before," or "No one will understand that application." This fallacy of noncompliance only confuses other users of the drawing (customers, suppliers, manufacturers, inspectors, etc.). Usually the only confusion that results is that the other users don't understand your misapplication. The rules in the governing documents of engineering are the same for both Government and commercial applications.

In this book I define the most common errors appearing on drawings. By careful attention to these areas of most probable mistakes, your drawings will be much more accurate and compliant to the established rules. I am not attempting to define how your drawings or designs are to be completed but I do direct you to the appropriate documents for this and the complete definition. The documents that I direct you to contain complete examples and definitions are the only appropriate governing documents. By reviewing the most common drawing errors and applying the proper applications will result in approximately 90 % of your errors being eliminated. Most are common sense and general drawing and dimensioning practices. Your complete understanding of what you have placed on a drawing and its proper interpretation is imperative as well as utilizing the proper drawing type for the intended application. Drawing types have not changed significantly over the years and most of the requirements have remained the same. Deviations from these requirements are always hard to justify as they have been developed after many years of application, discussion, and research. Basically all the rules are just common sense.

You must always temper the proper documentation requirements with the way it is preferred by your supervisor or company rules. Their views may not agree with proper documentation requirements and you need to present your input carefully. A change in company policy is never easy and should be approached with care.

Del Rio, TN, USA   Ron Hanifan

Contents

Chapter 1
Understanding Your Drawings and Their Requirements

In this book, I will explain only the most common drawing errors that appear on engineering drawings and the basic usage and understanding of the most frequently used drawings. By understanding and utilizing the proper types of drawings and reducing the most common drawing errors a much more comprehensive drawing package will result and a reduction in the drawing processing time should occur. Of course the more accurate the drawing will facilitate the inspection and result in more acceptable parts.

In the endeavor to reduce your engineering drawing errors it is imperative to understand the importance and the function of a drawing, how it is used, who owns the drawings and who the users are. The drawings that you prepare are very seldom yours, the program or the company that you work for. In many cases you're being paid to prepare these drawings by a customer, and your company is only a user of them. The rules for the preparation of drawings can vary with each contract.

Drawing and design requirements are usually the same whether prepared for a commercial program or a government contract. The rules for engineering documentation all originate from commercial ASME documents and this creates a commonality with all organizations, unless your company decided to deviate from these rules. These are the standards by which all drawings are made and they are universally understood.

The drawings you prepare may be just for research and development, modification of existing equipment, or formal production drawings. Don't guess; find out what your drawing and equipment contractual requirements are! You must know what the requirements and rules are if you are to reduce engineering drawing errors. Don't make the assumption that the drawing preparation requirements are the same as what you have done previously.

Drawings are used for the fabrication of parts but most importantly they are inspection documents as they specify end item requirements. Drawings should not specify methods of fabrication, assembly or inspection except when the method is essential. When a method is specified then it becomes binding on all users of the drawing.

© Springer International Publishing Switzerland 2015
R. Hanifan, *Perfecting Engineering and Technical Drawing*, SpringerBriefs in
Applied Sciences and Technology 139, DOI 10.1007/978-3-319-06983-8_1

If you allow your in-house departments to specify their requirements on the drawings then it may create conflicts with other users of the drawing. Your in-house departments and their requirements or applications may vary from those of the other users. Remember, these are only your in-house applications and they may not be pertinent to other users or your customer. Listing their requirements parenthetically is an acceptable method to convey this information.

In most cases a drawing belongs to whoever is funding (by contract) the design and documentation. That is your customer! You are only a user of the drawings that you prepared. You must become familiar with the contract and understand the requirements. The requirements may be your company's rules or customers. If there is a contractual requirement then the contract will specify design requirements, drawing requirements and delivery of items, drawings and any computer aided data. The contract may simply specify that your company's internal requirements are sufficient (hopefully you have published requirements).

Exercise caution when your drawings are prepared for a customer, as normally all the data that you specify on the drawings will belong to your customer. Be particularly careful about specifying in house processes, manufacturing processes, and test fixtures that you are not prepared to deliver. These items may be specified, but be sure to identify this information parenthetically, stating that it is for reference only. Never list this type of information on a drawing except when it is truly a reference for your organization only and isn't necessary for fabrication or inspection. If not listed as a reference it will do nothing but create warnings to your customer. He may require delivery of these processes or test fixtures. Your customer paid for these drawings and probably any special test fixtures that were developed using contract funding. Be very wary when specifying test fixtures as this frequently becomes a contractual and legal issue.

When drawings are to be delivered to a customer he will probably send these drawings to other subcontractors for quotes for manufacturing and reprocurement of parts or for spare parts. If you specify your in-house processes, they may conflict with how others may desire to manufacture these parts. Your in house processes are the way that you manufacture your parts and other users may not use the same methods. When stated as a firm requirement it does become inspectable information. If manufacturing wants to use a different method of manufacture, then the drawing must be revised, costing many hours of processing engineering changes and affecting other in-house departments. If your company is using outdated drawing delineation methods, don't be afraid to step out of the box and try something new. You undoubtedly will run into resistance from your in-house departments because change is always difficult to accept. Change, if it is really desired, must originate from the top and flow downwards.

Most customers are not as interested in the technical data that appears on the drawing but *normally* they are more concerned in the completeness of the Technical Data Package (TDP). The reasoning for this is because of the different engineering reviews that have been previously completed to verify that the parts and drawings agree. If the equipment functions as specified in the contract, then the equipment and drawings must be correct. It makes sense, if the equipment has been fabricated in accordance with the drawings (as verified by reviews) and the equipment functions per the contract, then logically the drawing content must be correct. The only concern would be for repeatability and reprocurement. At this juncture he is primarily concerned that he is receiving all the documents listed on the drawings. This becomes very difficult because of the many vendor callouts, manufacturers information, obsolete data, and in-house information that appears on drawings.

Always keep in mind that drawings are to define end item requirements and not impose a method of manufacture. For example, do not specify "drill and ream" for a hole but only specify the end item hole size (including any finishes unless otherwise specified). If you specify a method of manufacture, it will become an inspectable item, and when a part is received at your facilities, how will you verify that the hole was drilled and reamed?

It is imperative that all data on the drawing contains inspection criteria defining both acceptance and rejection criteria. When methods of manufacture are specified, such as a method of finish, fabrication, processing, or inspection then acceptance and rejection criteria shall also be specified. Methods are difficult to inspect and when drawings are delivered or reviewed by a customer, it causes even more problems as customers and their subcontractors methods may differ from those specified on the drawing. When a part being delivered from a subcontractor and it is ready for inspection, will all the data on your drawing be inspectable?

All drawings will contain errors, but you can eliminate many errors before the engineering design checker or your supervisor reviews your drawing. It will be processed much easier and you will look much better by eliminating the minor errors before the checker or your supervisor reviews it. A more concentrated effort can then be focused on the more important aspects of the drawing and design. You created this design and prepared the drawing and it is your reputation that is at stake! Remember your supervisor and the engineering design checker will see everyone's work and errors. They will know your capabilities and your weak areas. They know who produces good work and who doesn't. It is helpful to know what they look for, or should be looking for. Your drawing will speak for you and it is what others will use to measure you by.

Herein I will explain the most common errors on drawings and the most frequently used drawing types and uses. The data contained on these drawings, when not done properly, is subject to interpretation instead of being governed by universally published rules. Geometric dimensioning and tolerancing has become very complex, but it need not be. The best policy is to keep it simple, but stringent enough to build good parts. As drawings become more complex, the probability of misinterpretation increases. Also, it can increase the cost of the product and inspection tooling. Engineering rules are structured and uniform to allow for global repeatability in interpretation. These rules are the foundation of our profession and should never be violated, regardless how minor. The impact created from these violations may range from minor to major, immediate or future. Procurement, manufacture, inspection, stocking, and assembly and sparing of parts or equipment may be impaired. There normally is no bona fide reason to deviate from established and approved rules.

These rules have been time tested and are sound and logical. Usually the only real reason a company wants to deviate is that old company policies have never been updated. Frequently the excuse is that "we have always done it that way" or "our systems don't allow that." If your documents are reviewed by an outside source such as a government activity and you are written up for the deviation from the rules. You most assuredly will correct these deficiencies instead of facing the possibility of a loss of contract or be forced to request for a formal deviation to the contract. A simple violation, although not normally important, can be very costly. I have seen where just the placement of a sectional view (in violation of third angle orthographic projection) has resulted in a part being improperly made. Also not reading the contract that contained a requirement for the placement of a drawing note resulted in millions of dollars of cost. Upon receipt of a contract, identify those areas where you can't or don't comply and then submit a request for a deviation or a waiver.

Other outside sources will see your documents, whether they be customers, competitors, subcontractors or vendors. Be sure that you want your company to be judged by what they see.

Standard ASME rules not only pertain to geometric dimensioning, but to drawing preparation, revision, abbreviations, inspection information, drawing format and all information contained on the drawing. Always understand every line, dimension, word and sentence that you place on a drawing. Most of these requirements are contained in your engineering procedures manual, contract, ASME Y14.100, and in sub-tier documents. If you don't have an engineering procedures manual, then you should develop one for standardization and uniformity in interpretation. Most ASME/ANSI documents are not logically intended to be fully complied with and in your engineering procedures manual (and in the contract) are where you need to list your deviations. When this is done, then there are no surprises to anyone.

The words you use on a drawing are very important. If you use a word such as "should" or "may" and your intent was to make it a mandatory requirement, then you should have used the word "shall. There is always confusion with the words, vendor, supplier, customer, fabricator and design activity. I recommend that you become familiar with these words because with the misapplication of these words can cost millions of dollars of unnecessary cost or omission of inspection/testing. The wording in notes is very important so that they convey the proper intent. Occasionally the intent is buried in superfluous wording which hinders the true intent. Always use the simplest of words and properly construct the sentences to convey the intended meaning.

Because of the complexity of Geometric Dimensioning and Tolerancing (GD&T), millions of dollars are spent yearly teaching the interpretation or revisions, and it has become a huge business. There are classes, videos, seminars, books, etc that are available, and many are offered from the same sources that are on the ASME committee that created the complexities of GD&T guidelines. With every update to ANSI/ASME Y14.5 new reviews and education begins all over again. It would be nice if these committees would revise ANSI/ASME Y14.5 to create simplification instead of making it more complex. I have used geometric dimensioning and tolerancing for over 50 years and I have never seen a simplification of it, only more complex usages, that are very rarely used and that obscure interpretations. The understanding and application of GD&T has not improved since I reviewed drawings when it was first applied. I see the same errors and misapplications that were prevalent then. With over 50 years of GD&T it would seem that the application of it would improve. I do recommend that you take a basic course of GD&T before attempting to use the complex applications.

Chapter 2
First Step Is Done, Design Is Complete

You have labored many hours developing your design, and now you are finally done. Whichever computer aided design (CAD) system you used, it has been proven to be a tremendous asset, and you are very proud of your accomplishment. However, now it is time to prepare drawings so that parts can be fabricated, inspected, subcontracted, assembled, tested, or procured. The initial design is a very important part of your effort, but now it is time for the real world and the documentation of your product. This is the most important step in your design. This is what everyone will see, and it is what people will use to evaluate you and your accomplishments by. The drawing is the one item that is visible to everyone. The primary focus is to document your design on drawings with such accuracy that no bad parts will be made and to ensure that there is no ambiguous information. The reality is that every drawing will contain errors; some will be serious and some minor. Even seemingly minor mistakes can cost millions of dollars. Regardless of the type of error, someone will be impacted.

From your drawing, procurement will order materials, drawings will be sent off for parts to be fabricated, production will set in motion the assembly steps, inspection plans will be developed, and drawings will be sent out to prospective manufacturers. All drawings will contain mistakes, but with careful attention, the severity of these mistakes can be minimized. Everyone will scrutinize your drawing, and your reputation, your department, and the company is at risk.

What you designed, as shown in your CAD system, is designed at nominal (average), and everything fits. It is perfect! But in reality, when parts are fabricated, they will contain many errors because of tool wear, operator errors, or most importantly "misinterpretations". The drawing that you have prepared defines how much error is acceptable.

The data on your CAD tube normally goes nowhere; it is the paper (drawing) that will now receive everyone's attention. Manufacturing, procurement, and inspection will be the main users of the data that you have developed. Drawings will be sent out for quotes to fabricate or buy, and parts will be inspected, manufactured, and assembled from the information contained on these drawings.

© Springer International Publishing Switzerland 2015   7
R. Hanifan, *Perfecting Engineering and Technical Drawing*, SpringerBriefs in
Applied Sciences and Technology 139, DOI 10.1007/978-3-319-06983-8_2

The information you place on drawings is to be presented in such a manner that there can be no misinterpretation of the intent of the end item. All features and information require complete inspection criteria (acceptance/rejection). This sounds easy; however, it isn't.

Each and every part of your design has to be documented in some manner. It may consist of new drawings or usage of existing drawings, commercial specifications, vendor parts, and so on. Materials and finishes need to be specified, configuration completely defined, and any other necessary characteristics such as installation of inserts, burr removal, flash specified, parting lines, surface texture, and so forth all need to be specified.

Requirements for delineation vary, depending on the type of item being depicted. It is very important to keep in mind the intended uses of the drawing. It may have to meet requirements imposed by a contract or your own company's standards. The drawing also becomes a legal and binding contract between your company and other users. Primarily, your drawing will be used by a manufacturer, but it is not a manufacturing drawing. It is an end item drawing, defining only end item requirements and not the methods of accomplishing the action. Hence, only a hole size is given, and not the method of manufacture, unless the actual method of manufacture is critical. Remember a method is very difficult to inspect, and the method, when specified, does become an item requiring inspection.

You have dimensioned your parts completely, and when done properly, inspection will have no problem understanding your intent. In all probability, you used geometric dimensioning and tolerancing (GD&T) to define your product. GD&T is a very complex endeavor, and you must understand exactly what you have placed on your drawing. Understanding of your tolerance zone, the shape and extent of it, and how the size of it may increase/decrease is of the utmost importance. The more exotic method used in your application of GD&T then the greater the risk and difficulty in the interpretation. If you don't understand completely the tolerance zone, then you can't expect others to understand. If the application is not contained in ANSI/ASME Y14.5, then you cannot use it. ANSI/ASME Y14.5 *contains the only approved rules*, and any other documents contain only someone else's interpretations. The best rule of thumb is to keep it simple but stringent enough to build good parts. Remember, tight tolerances increase the difficulty of fabrication and inspection, all of which impacts the cost.

Usage of GD&T should not increase the cost of products because when used properly it can loosen tolerances. Frequently I have heard that subcontractors don't understand GD&T and that is used as justification for not using it on drawings. The reality is if you don't use GD&T then there is no interpretation of coordinate dimensioning. Frequently subcontractors don't understand the GD&T you placed on drawings and that is because usually because of your misapplications of GD&T.

It is very important to pick the right type of drawing for the intended function. There are detail (part) drawings, assembly drawings, testing specifications, many different types of procurement drawings, drawings for altering existing parts, reference drawings (such as piping documents), schematics, interconnect diagrams, installation drawings, and others. Assembly drawings will contain a Parts List or

Bill of Material (separate or integral). The most common of these drawings, and their requirements, is discussed in Chap. 4. These drawings and their requirements are controlled by ASME Y14.24 and ASME Y14.100. ASME Y14.24 specifies the types of drawings and their requirements, whereas ASME Y.14.100 specifies approved engineering drawing practices.

Chapter 3
Understand the Data That You Have Placed on the Drawing

Drawings and the errors that appear on drawings haven't really changed in over 50 years. The only thing that has changed is the manner of drawing preparation and the dimensioning practices. *Dimensioning and tolerancing* has been expanded by use of a more advanced *geometric dimensioning and tolerancing* to clarify the intent and to expand on applications. Geometric dimensioning and tolerancing has been around and widely used since 1960s, but the applications were much simpler and easier to understand back then. The basic applications and interpretations have not changed. It is imperative that the designer understand what he has invoked on the drawing, but all users of the drawing must also understand completely the intent.

The errors that appear on drawings today are the same as they always have been, except now CAD and modeling have added new types of problems. Previously the dissemination of drawings and information was slow and was comprised of paper copies that were transmitted by mail services which provided time for reviews, however now with Computer Aided Design the information is transmitted instantly and frequently even prior to any competent reviews.

In the usage of GD&T, the primary mistake that always persists is dimensioning from a centerline of a feature without an angular orientation (clocking) datum, dimensioning from an undefined centerline, and the listing of processing and manufacturing information. On a drawing, dimensioning from a centerline is so easy because the crosshair provides a horizontal and/or vertical orientation. In reality, this crosshair doesn't exist. It is normally an axis. This same problem has existed since the beginning of GD&T.

The engineering drawing is used by many people, and it is used primarily as an *inspection document* and a *legal document*. The drawing forms an agreement between the procuring activity (design activity) and a vendor, manufacturer, or subcontractor. The drawing completely defines the end item requirements of the item. In the endeavor to completely define an item, occasional errors or ambiguities occur that obscure the intended interpretation. These mistakes in interpretation can lead to rejected parts and legal complications. If the rules of ANSI/ASME Y14.5

© Springer International Publishing Switzerland 2015   11
R. Hanifan, *Perfecting Engineering and Technical Drawing*, SpringerBriefs in
Applied Sciences and Technology 139, DOI 10.1007/978-3-319-06983-8_3

are followed, then everyone will have the same interpretation. With proper application of simple GD&T and end item documentation, most of these misinterpretations can be avoided.

Manufacturing from model and automatic checking by CAD system: There are instances that a model is sent to a subcontractor in lieu of a drawing; however, this very seldom works unless stringent rules are invoked and followed. Advances are being made whereas models are used; however paper copies are still frequently used by inspectors, in-house departments, vendors and fabricators. Many in-house departments do not have the capability or assets to access these CAD programs. When models are used for fabrication and inspection it is important that they be error free. Modeling continuity between designers is a must but very difficult to enforce. When you provide a model you do incur liability as to the accuracy of their content so it is very important that they be error free and interpretation consistent. One of the major problems when utilizing a model for fabrication is that it is assumed that models are prepared at nominal finished size and that is not necessarily true. Fillets and rounds may or may not be shown, holes are not shown at nominal size, example a hole .250 diameter +.006 −.001will be shown at .250 not .2535 diameter which is the true nominal size. Features that do not have even tolerances are not shown at their true nominal size. Example a feature that contains a tolerance such as +.010 −.000 will not be shown at its true nominal size. Also models may not contain considerations of platings or coatings. A coating thickness may be very thick and if the thickness is included in the model it may cause a problem for fabricator as they may be manufacturing from the model are not necessarily the true manufacturing dimensions. If the model does not included the coating thickness then interference checks cannot accurately be performed.

It is very important that *the drawing that invokes the model* (**RDD**, **R**educed **D**imensioned **D**rawing, Undimensioned Drawing are the most common terminologies) that clarification be specified as to:

Sizes and tolerances of fillets and rounds (they are either shown or not shown in the
    model and if not shown specify their sizes).
Notes pertaining to breaking of sharp edges.
Dimensions that do not have equal tolerances.
Note defining governing tolerance, usually a profile tolerance.
Interpretation of dimensions of finishes and coatings.
Surfaces textures.
Model version identification.
Features requiring special considerations such as flatness, roundness, etc.
Datum identification.

Software programming has not been developed yet to avoid these problems and interferences checks are compromised by some of the above outlined problems.

The information you place on drawings is to be presented in such a manner that there can be no misinterpretation of the intent of the end item. All features and information require complete inspection criteria (acceptance/rejection). This sounds easy; however, it isn't.

The main focus of GD&T is to remove any areas of misinterpretation and to allow looser tolerances. That sounds simple, but the opposite has occurred. ANSI/ASME Y14.5 has many extreme examples and definition of usages that most of us will never apply. It's a shame that so much of ANSI/ASME Y14.5 is devoted to conditions that are very seldom used. The best thing to do is to keep it simple. Normally, positional tolerances, profile, flatness, perpendicularity, runout, and coplanarity are enough to define an item. Excessive use of datums needs to be avoided, along with the inappropriate use of modifiers. The simpler it is, the less prone it will be to misinterpretation. If you don't understand all the presented data and tolerance zones, then it is very likely others will not understand it either. Always understand all tolerance zones that you have shown, and ensure that all data contains acceptance and rejection criteria.

Always present information in the affirmative manner. Remember "shall" indicates something mandatory, and one cannot inspect to data that is presented with "may," "should," or "optional." When information is presented with a "may," "should," or "optional," then it might as well not even appear on the drawing. Example: *"Corners should have an .04 Radius"* *"Corners shall have a minimum Radius."* Data presented in this manner cannot be inspected.

All information presented on a drawing shall be capable of being inspected. Never present data that may be vague, ambiguous, or containing a methodology. Review what you listed on the drawing and ensure that it all data contains inspection criteria. There is always the desire is to control the method of manufacture or assembly, but this usually comes with adverse affects. When one introduces a method, then the end result and the method becomes a requirement for inspection. Much of this data is more appropriately shown on in-house processes. A manufacturing process is much easier to revise than every drawing where the data is shown.

With attention to the following outlined areas, many errors will be eliminated. The following are examples of the most common problems appearing on drawings.

The Most Common Error

Dimensions Originating from the Center of a Feature

1. Center of a Feature
A center of a cylindrical object is an axis, not a crosshair. The presentation on a drawing is very misleading and the origin of most misinterpretations. Information originating from an axis has no real orientation unless another feature is utilized (datum) to establish angular orientation (clocking). *Mentally remove the crosshair and visualize an axis.* Another method that I encourage to help understand the axis is instead of drawing as a crosshair (plus sign) is to draw the crosshair rotated 45° as an **X** or use a dot, then dimension from the center of

the crosshair or dot as before. This simplification helps in the interpretation of an axis.

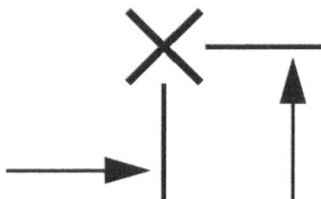

2. Dimensions from the Center of a Cylindrical Feature

It is very probable that most dimensions originating from a centerline will be displayed incorrectly if GD&T isn't used. Use a datum to clarify the origin of the centerline. Normally, the centerline is undefined without datums and GD&T or there is no angular orientation (clocking). Many times the requirement (rule i of the Fundamental Rules of ANSI/ASME Y14.5) is misinterpreted that specifies "A _90° angle applies where center lines_ and lines depicting features are shown on a drawing at right angles and no angle is specified." It is thought that the extension lines/dimension lines coming from the center of an axis indicate that features are 90° apart. It is imperative to remember that these lines are extension or dimension lines only and the axis is in a completely different plane. Never confuse these lines with an axis. A feature dimensioned from an axis, even though it is dimensioned horizontally and vertically can still rotate if no tertiary datum is specified to control angular orientation and still be within the dimensions specified.

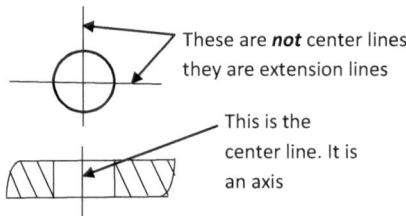

These are **not** center lines, they are extension lines

This is the center line. It is an axis

3. Identification of Feature

Always identify features by using GD&T (a datum) to define the centerline from which dimensions are originating. Frequently features are located from an unknown centerline. Identify the feature that the centerline is originating from. Never just show a centerline without defining it and its origin.

4. Angular Orientation

Usage of a tertiary datum in the feature control frame is probably required (if location is to be controlled) to establish angular orientation of any feature located from an axis. See paragraph 2.

5. Notes

Notes are improperly written, ambiguous and lacking acceptance/rejection criteria. See Chap. 5 for note requirements.

6. Feature Control Frame

Use the modifier MMC (i.e., maximum material condition) or LMC (least material condition) if applicable when an additional bonus tolerance is desired or appropriate. Always understand if it is beneficial for this additional tolerance and is it logical. Is there a relationship between the size and the bonus tolerance? Secondary and tertiary datum callouts list MMC or LMC only when that *additional bonus tolerance is actually desired.* If no modifier is specified then, by omission, RFS (regardless of feature size) applies. Do not use the modifier MMC as a default. Always fully understand what you have imposed on the drawing.

Other Frequent Errors

7. Tolerance Zone

Always understand the shape and location of the tolerance zone for each feature and how any additional tolerance (MMC or LMC [maximum material condition or least material condition]) affects it. All features require both a size and a locational tolerance. Use coordinate dimensioning (plus or minus) where possible and GD&T where coordinate dimensioning is not sufficiently accurate or the intent is ambiguous. The problem with coordinate dimensioning is that there is usually no reference to datums and perpendicularity of surfaces is not established. Additionally, there is no definition of interpretation of coordinate dimensioning.

8. Datums

Always select functional datums. Never use a datum where the feature is inaccessible or of insufficient size for contact. When the datum cannot be contacted because of size or otherwise being inaccessible, then the datum is useless. Care should be taken not to create an excessive number of datums. Normally, three or four are sufficient. Frequently, corner or fillet radii can minimize the actual contactable surface of a datum surface. A datum surface must be accessible. *Example:* Frequently, a mounting hole where a pin is installed is utilized as a datum. If a pin is in the hole, then the hole is not accessible. A better application would be to use an exposed feature on the pin. Remember inspection tooling is required to simulate these datums. Normally the largest surface is the primary datum and it normally controls perpendicularity.

Castings

Select datums such that the feature will not be machined and removed at the next level. On the machining drawing of the casting, locate the primary machining datums from the cast datums. If datum points are utilized, this will restrict your application of datums. If datum B is established by two points and datum C is one established point, then they should not be used as a primary datum, and datum C should not be used as a primary or secondary datum.

Modifiers: MMC, LMC, or RFS

Never use MMC as a default. Commonly, MMC is used without any understanding of the consequences. Never use a modifier unless you completely understand how this will affect your tolerance zone and only if you want this effect. Frequently, actual locational tolerance zones can double or triple with inappropriate use of MMC. Overuse of MMC or LMC, without a full understanding, will only compli-cate the interpretations and compromise your design. It can be very embarrassing if you are asked to explain the exact interpretation to a manufacturer or inspector, and provide the appropriate reference in ASME Y14.5 of your application. It can be very awkward if you can't find the rule or you don't know the interpretation. Remember, this data affects acceptance and/or rejection of parts.

9. Feature Control Frames
If no modifier is shown (MMC or LMC), regardless of feature size (RFS) applies. Previously the symbol for RFS was the letter "S" enclosed in a circle. There no longer is a symbol for RFS, and it applies by omission. If you forgot to place a modifier in your feature control frame, then by omission the feature is RFS (regardless of feature size and no additional tolerance is applied). The question will always arise; did you intentionally omit a modifier?

10. Tolerance Zone
In a feature control frame for positional tolerance, if you do not show a diameter symbol, then the tolerance zone is not round. This will always create confusion, especially if the application or lack of the symbol is not logical. It will not be clear if you did not show the diameter symbol intentionally. Always review you feature control frame to verify if all the symbology is correct, datums and modifiers shown properly. Even though this appears to be a very simple item,

the processing of drawing revisions and all the consequences (inspections work stoppages, etc.) can cost thousands of dollars.

11. Basic Dimension

Confusion occurs in the interpretation of a basic dimension. It only describes the theoretical exact size, location, profile or location of a feature or datum target. By itself, it does nothing and it needs to be accompanied with a feature control frame that identifies the accompanying tolerance.

12. Changes That Affect Interchangeability

Be extremely cautious when revising drawings where a part is changed to the extent that interchangeability is affected. If performance or durability is affected to such an extent that the previous version must be discarded or modified for reasons of safety or malfunction then a new part identifying number (PIN) shall be assigned. This change can also flow into next higher assemblies up to the level in which interchangeability is reestablished. Any change where an old part and the new part are not completely interchangeable requires that a new part identification number (PIN) be assigned.

13. Reference Designations

Reference designations are often assigned inappropriately, not shown in the Parts List, not assigned and the Reference Designation marking is not shown or misapplied. See Chap. 6.

14. End Item

Drawings shall only define end item requirements. Show only the end item result, not the method of obtaining the result. When a method is expressed, then it becomes an inspection requirement for verification, and acceptance and rejection criteria is required. An engineering drawing is a legal document and an inspection document, not a fabrication drawing, although at times a method of fabrication/assembly may be specified, when that action is critical. Do not confuse critical with "that's the way manufacturing wants it" or "that's the way we are going to make it."

Drawings should never list processes, manufacturing data, quality assurance, or environmental information except in those cases when this information is essential to the definition of the engineering requirements. Whether drawings are prepared for Government contracts or commercial programs, it is a good business practice *not* to provide this information. Your internal processes should be kept proprietary.

When unnecessary information, manufacturing, quality assurance, or processing data is listed on a drawing, or a drawing is structured to mimic assembly or inspection sequences, it imposes a burden on other users of the document (customers or other manufacturers), as it does not allow them the freedom to utilize their normal fabrication, inspection, or procurement and/or subcontractor operations without revision to the drawing. All information that is listed on a drawing becomes a mandatory requirement and requires verification of compliance unless it is specified as reference or nonmandatory.

Drawings are prepared to support competitive reprocurement and maintenance for items substantially identical to and interchangeable with the original items. The drawing shall provide the design disclosure information necessary to enable and maintain quality control of items so that the resulting physical and performance characteristics duplicate those of the original design. The drawing shall provide the necessary design, manufacturing, and quality assurance requirements information necessary to enable the reprocurement or manufacture of an interchangeable item that duplicates the physical and performance characteristics of the original product, without additional design engineering effort or recourse to the original design activity.

Drawings should only disclose details of unique processes, not published or generally available to industry, when essential to design and manufacture.

In some cases, processing information may be shown, but that data should be prefixed with an appropriate note such as "*nonmandatory, manufacturing information.*" It is imperative to remember that all data on a drawing requires verification of compliance. Information such as *drill and ream, clean, grind, broach, #10 drill,* and *polish,* when not prefixed with this note, it becomes a mandatory requirement. Evaluate the need to disclose processing information, as it could become part rejection criteria.

Specifying any unnecessary processing, quality assurance, and manufacturing data on drawings should be avoided except when it is essential to describe the item. The following are some of the major areas that may be affected by the disclosure of this information.

a. Inadvertent release of information to sources (competitors) outside of your company, pertaining to your manufacturing processes and sequences of assembly. This information should be released only when it is essential to the design or assembly. This information should appear in manufacturing processes, which would be for internal purposes only and not available to competitors.

b. Inspection of an item becomes difficult when a process or a sequence is invoked instead of verifiable end item requirements. End item requirements, when specified properly, have measurable parameters that are to be verified. However, when a process or sequence is specified, the inspection or verification is difficult, as there normally are no measurable parameters specified. Usable parts may be rejected for failure to comply with a method or a process, and the recovery or rework becomes ambiguous.

c. The listing of quality assurance or inspection information criteria, when not contractually required, places a burden on any other users of the drawing and invokes a mandatory requirement on the current user that may not be deviated from, without a revision to the drawing. These requirements are normally only in-house procedural requirements and are best placed on in-house documentation such as inspection plans, purchase orders, and manufacturing data sheets. This will allow changes to be made without costly changes to the drawings and will not disclose to other users of the drawings your internal functions and methods of inspection.

15. Acceptance/Rejection Criteria

Evaluate all data on the drawing, especially the information contained in the notes. Does all data have inspection *acceptance and rejection* criteria? Have you placed redundant information on the drawing that creates confusion to the interpretation? Evaluate this data as if you were the inspector of the part and how you would inspect and verify the requirements on the drawing.

16. Using a Profile Tolerance vs. Positional Tolerance

A profile tolerance is usually used to define curved or irregular surfaces. In many cases, a profile tolerance is desirable in lieu of a positional tolerance (especially in locating radii), and each case should be evaluated to verify which is more functional. *Positional tolerance* locates the center of the feature, while *profile* locates the surface.

17. Is Every Dimension Shown?

The most important item is to verify that all dimensions are present with a tolerance and that they are presented in such a manner as to be subject to only one interpretation. Missing dimensions are easily overlooked because of CAD systems utilized in drawing preparations.

18. Fillets and Rounds

In models and on the drawings, most features are shown as sharp, with a note defining that all fillets and corner radii shall be XXX (a specified radius). This is very misleading, as many small features, when these fillet or corner radii are applied, will be nonexistent or rounded. Extreme care needs to be taken when specifying fillets and corner radii. Visualize the effect of fillets and rounds on features shown square in this part (Fig. 3.1).

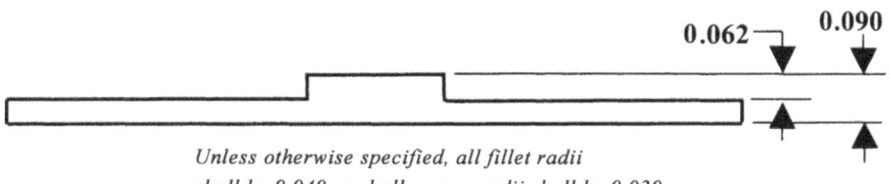

Unless otherwise specified, all fillet radii
shall be 0.040, and all corner radii shall be 0.030.

Fig. 3.1

19. Stock Thickness

Stock thicknesses should never be listed. A stock thickness is the thickness that the part begins with or the thickness that is supplied, not the finished thickness. It is recommended that the material thickness be shown with an appropriate tolerance in a view or note. I have seen cases involving expensive formed parts that called out a stock thickness that could not be rejected because of a material thickness not being acceptable.

20. Coatings and Platings

Clarify if tolerances apply *"after platings and prior to coatings."* A coating is normally a primer or paint. Also, these coatings are known as *organic coatings*. Some coatings can be very thick and require that the part by fabricated undersize to accommodate these coatings if it is stated that the tolerances apply after coating.

21. Dimensional Refinement

When dimensions are used as a refinement, such as flatness, parallelism, roundness, and so forth. Ensure that these dimensions actually are a refinement to the size tolerance and that the degree of refinement is sufficient for the cost and is a necessity of the design.

22. Nominal

Avoid usage of this term; there is no definition of the word *nominal*. Although in engineering it is commonly understood to define a mid range to tolerances, it is not a defined word and is subject to interpretation.

23. Required vs. Places

These have different meanings. When stating "required," the term means that, at that particular point of application, the specified quantity is required. For example, "A washer, two required" means that, at that particular place, there are "two" washers required (or installed). "Two places" means that "one" washer is used in two different locations. Most companies no longer use the word "places"; they just use the letter "X." Instead of stating "two places," they use "2X" although at times the proper application of the terms can clarify intentions.

24. Drawings Only Depict Perfect Parts

A drawing defines a perfect part, not a broken, dirty, or out-of-tolerance part, and the information placed on the drawing should define it accordingly. Don't specify information in anticipation of imperfect parts being manufactured. Of course imperfections will occur but deviations should be shown on manufacturing or inspection processes, not on the drawing. The following are some examples of information commonly placed on drawings that anticipates a part being manufactured improperly. This type of information belongs elsewhere such as manufacturing processes or purchase orders. Examples include:

- Grease and dirt shall be removed and part solvent cleaned
- Grind to achieve surface texture
- Touch up paint

These are processes, and they assume that the part is contaminated with grease or dirt, that the paint is chipped, or that the method to achieve the end result must be specified. These requirements cannot be met if the part is not contaminated or the part is not ground. Specified in this manner, the method requires verification in addition to the end result. These are manufacturing or suppliers problems and should be addressed on their internal processes, not on the drawing. Remember, these parts/drawings may be used by other design activities, and they may not have your manufacturing problems.

25. Revisions/Drawing Changes

Simplify your revisions. All CAD drawings are redrawn whenever they are changed. With appropriately written change notices (ECN/NOR/etc.), the incorporations can be simplified. The days of change balloons and lengthily written revision columns went away 50 years ago. I strongly recommend that if your activity writes a complete revision history on the drawing, stop this practice and place the entire change information on the drawing change notification. There is no longer a need to write a complete description in the revision block. When contractually permitted, state only the change authority, such as *Revision*: *A*, *ECN 12345*.

26. What Requirements Does Your Drawing Have to Meet?

All drawings have to meet some requirements. It may only be company internal requirements or contractual requirements by a customer. Don't just assume that the drawing requirements are the same as the requirements of the last program you were on. *Check it out!* As a minimum, there will be internal requirements (i.e., company requirements). Frequently, requirements of programs will vary, even within the same company. In all probability, your drawings will be sent outside of your company to buy or fabricate parts, so it is imperative that they provide uniform interpretation. **The drawing is the book cover of your** company as well as an indication of your abilities. Remember others will judge you and your company by the drawing and the errors on it.

If the drawing is made in accordance with a Government contract, find the Contract Data Requirements List (CDRL), the appropriate Data Item Description (DID), and any other appropriate information. This will define exactly your requirements for drawing preparation and delivery requirements as well as engineering information. Frequently, the company will extract these requirements and publish them so that they are available to all. They might appear in a Statement of Work (SOW) or another similar document. Regardless of the requirements (company or Government), the intended use is always the same, and you want to provide a universal interpretation. Basically, drawing requirements are all the same, whether Government or commercial, and they originate from ASME documents for their guidelines. (Reference ASME Y14.24, ASME

Y14.100, and other sub-tier documents.) The customer (Government) is required to prepare the appropriate documentation to specify the Technical Data Package (TDP) requirements for that particular contract. He may want delivery of the drawings, no usage of vendor part numbers, his drawing formats, various associated lists, and so on. Most importantly, always verify what the customer expects and what your company has agreed to supply.

27. Weight

Often, it is believed there is a need to list the weight on a drawing. This can be a very expensive item and nearly impossible to maintain accurately. I recommend that it be placed on the drawing only if required contractually. When not contractually required, list weight as a "reference item" and show it only as a "maximum." Weight is really not a problem, except temporarily. Once the weight requirements are met, it is no longer important. If not contractually required, no value is added to the design to list the weight. Weight can be derived from the CAD model.

The problem with weight is that it is usually accurate only until the first design change. When a weight is changed, the drawing above (next higher assembly) is also affected. Then, if that assembly is affected, then its next assembly also is affected, and so forth until reaching the very top level. We know that management will never sponsor revising every drawing in that particular chain every time a simple design change is made.

When weight is not listed as a reference item, then it becomes an inspectable item, even if it is listed only as a "maximum weight." The problem with the weight being listed as a maximum is that it was probably calculated at nominal values and without finishes, fillets, and rounds. Unless there is a dire need or a contractual requirement, do not list the weight, as it becomes an inspectable item, which can result in rejection of parts. If a part is rejected, rework of a good part becomes nearly impossible, and it was probably just an error in the initial weight calculation.

28. Undisclosed Test Fixtures and Information or Missing Testing Data

Avoid reference to undisclosed test fixtures and digital data (software) information. This occurs commonly on drawings, such as test requirements that make reference to a test fixture or digital data. When this information is listed on a drawing it is required to specify the part number or drawing where this information is contained. This is always a prime area of concern to any customer and always the focus of much of his attention. Without disclosure of this information, no one can repeat this test without recourse to the design activity. When this is only internal information, it should be so noted. Also, when the customer is paying for the drawings and equipment he will also want the test fixtures or software that was developed with his funding.

Another prime area of concern for customers is when electrical items are lacking electrical testing information. Double check drawings of electrical items, especially printed circuit board assemblies or electronic assemblies to ensure that appropriate testing data is present.

Chapter 4
Dimensioning and Tolerancing Errors

This chapter provides examples of many of the common dimensioning and tolerancing errors appearing on drawings.

Most misapplications of *geometric dimensioning and tolerancing* (GD&T) exist because of improper usage of an *axis*, no angular orientation (no clocking feature), or misapplication and lack of understanding of the implications of using maximum material condition (MMC). The usage of the modifier MMC on *datum features* is commonly used as a default, with no knowledge or concerns of the affects. It would be better to use RFS as the default, as there are no additional tolerances incurred, and then there is no affect. The common reason that MMC is used on datum features is that it provides additional bonus tolerances. If you don't know how these additional tolerances will affect your part and dimensioning scheme, then it is best to stay away from them (see Fig. 4.1).

© Springer International Publishing Switzerland 2015
R. Hanifan, *Perfecting Engineering and Technical Drawing*, SpringerBriefs in Applied Sciences and Technology 139, DOI 10.1007/978-3-319-06983-8_4

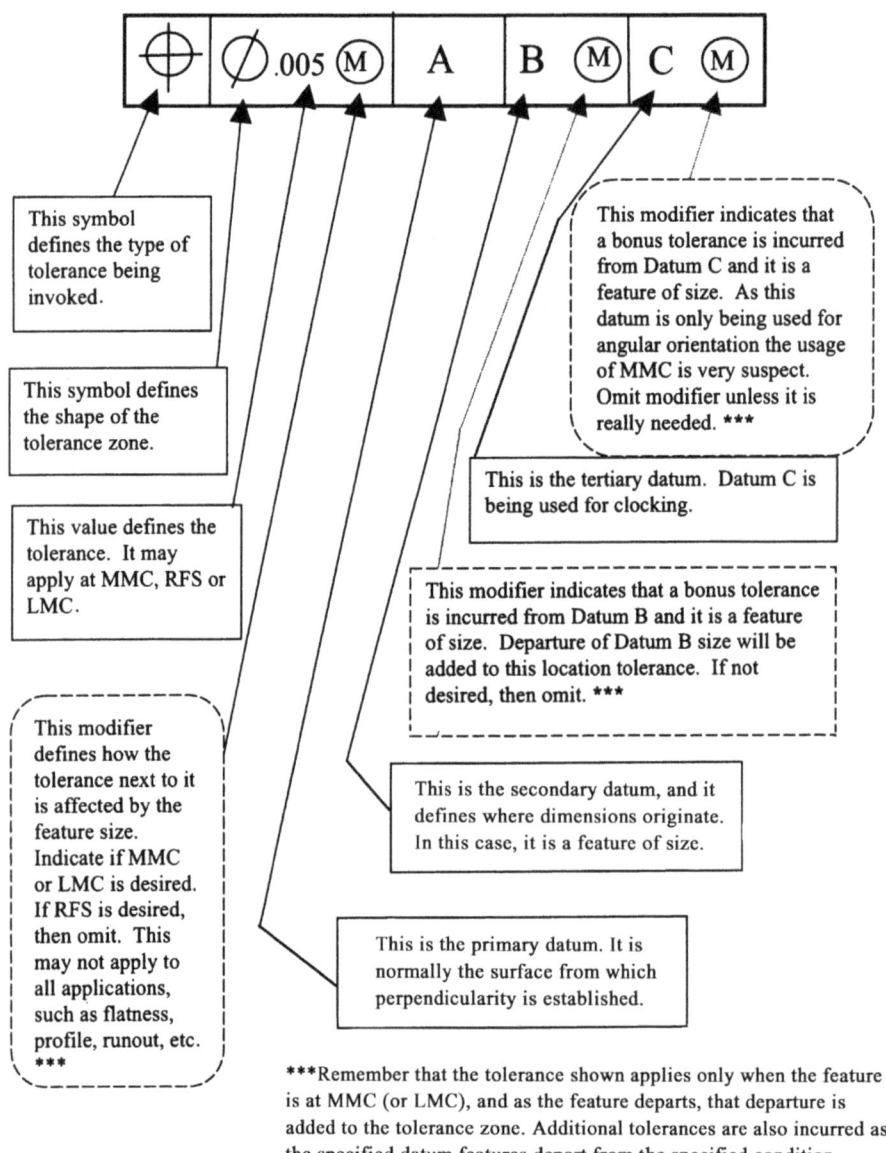

Fig. 4.1 Feature control frame

The center of a round feature is not a crosshair, *it is an axis* and it extends through the part. Round features are commonly used as datums. Centerlines (crosshairs) are always shown, but it is the axis that is the datum (see Fig. 4.2).

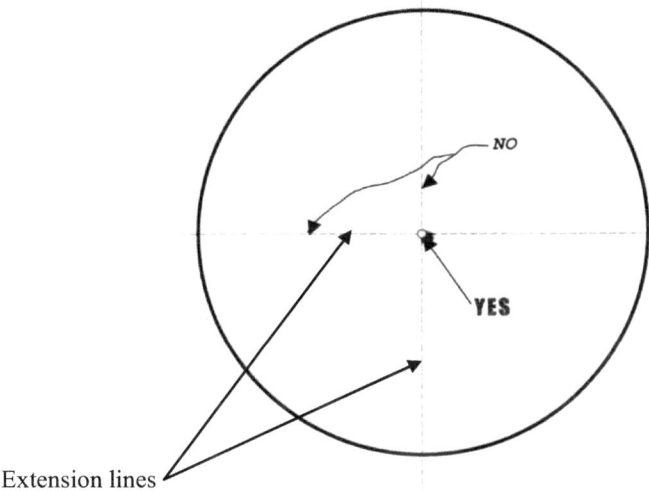

Fig. 4.2 The center of a round feature is not a crosshair, it is an axis

**Axis, but the
axis of which
feature is
not defined**

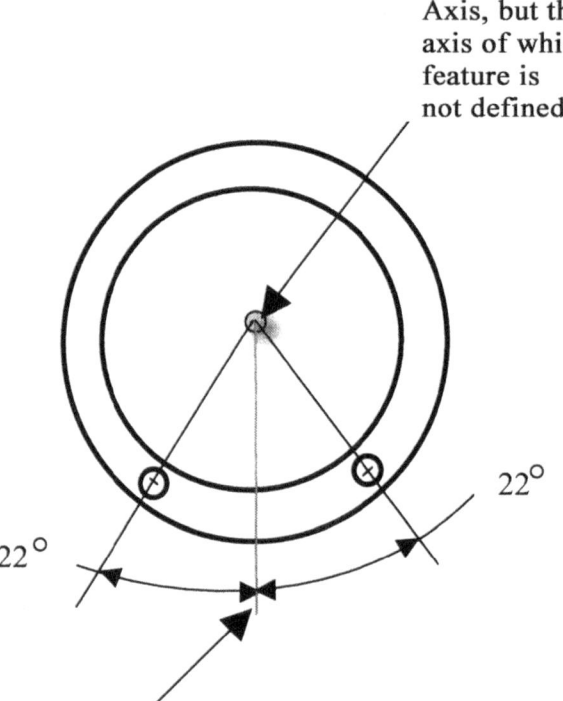

22°

22°

22°

**To define an angle originating from here (undefined
centerline) is rather meaningless. In actuality, it can be
any angle. The angle between the two holes is what is
critical, and in this case no angular orientation is
necessary.**

Fig. 4.3 Dimensioning from a centerline

This is a typical example where there is no definition of origin of the center lines
of angles. Use GD&T to define the angular orientation of the holes (when required)
and which feature the holes are located from. An angular dimension from the center
of the diameter to the holes is unimportant. The angular dimension of hole to hole is
normally more critical (Fig. 4.3).

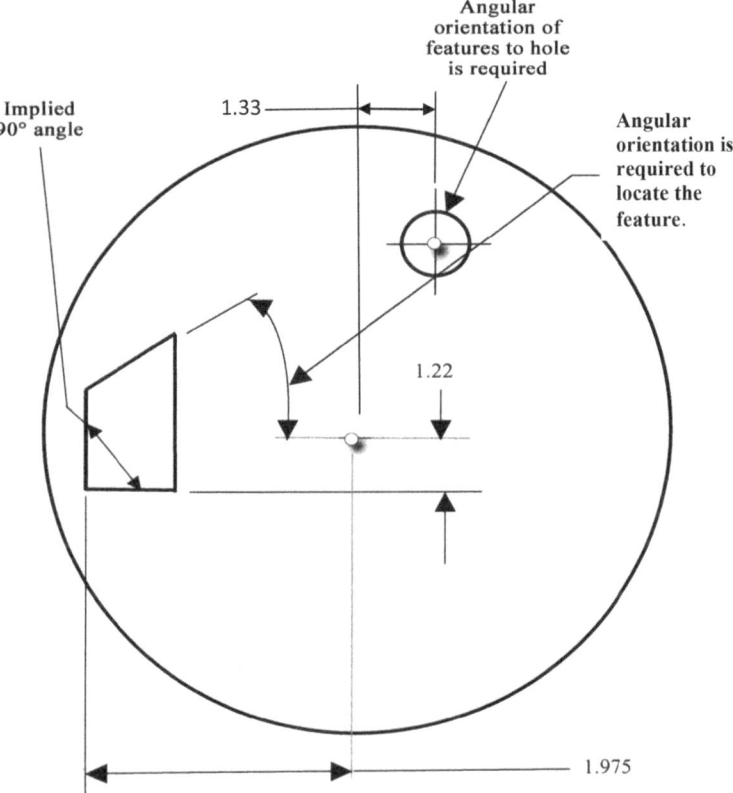

Fig. 4.4 Undefined centerline

Use GD&T (basic dimensions, datums, profile and positional tolerances) to define the internal features with respect to each other. As stated and shown, there is no angular orientation (Fig. 4.4).

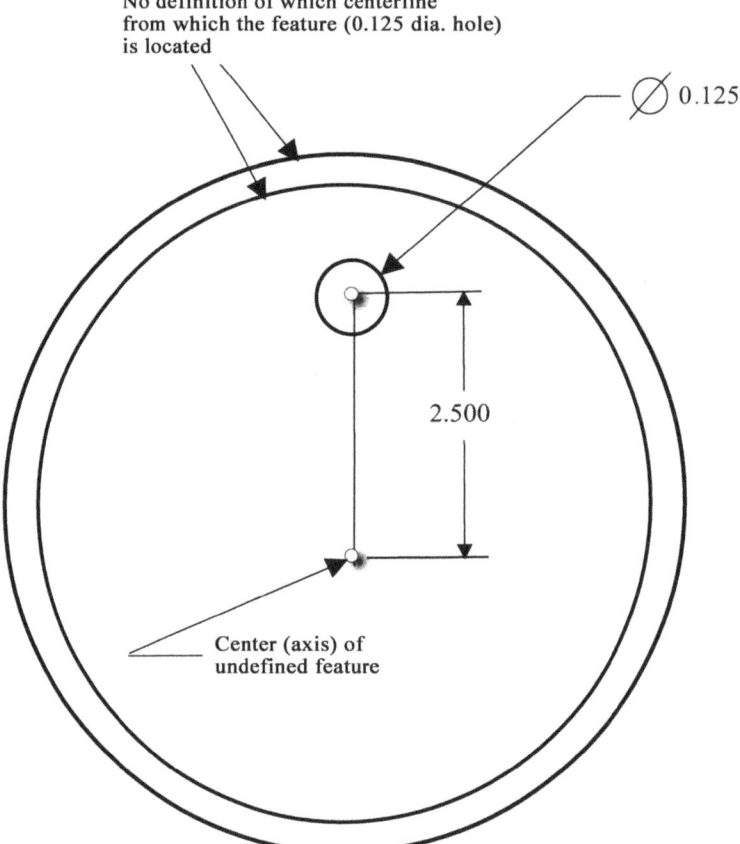

Fig. 4.5 Clarifying centerline

Use GD&T to define which center line the hole is located from. Also, there is no location of the two large diameters with respect to one another (Fig. 4.5).

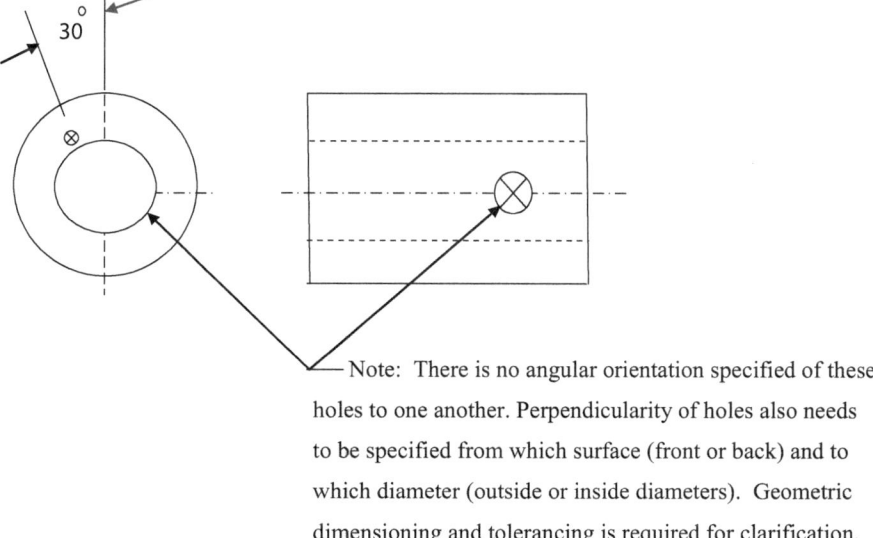

Note: There is no angular orientation specified of these holes to one another. Perpendicularity of holes also needs to be specified from which surface (front or back) and to which diameter (outside or inside diameters). Geometric dimensioning and tolerancing is required for clarification.

Fig. 4.6 Angular orientation

Note: There is no angular orientation specified of these holes to one another. Perpendicularity of holes also needs to be specified from which surface (front or back) and to which diameter (outside or inside diameters). Geometric dimensioning and tolerancing is required for clarification.

GD&T is required to establish angular orientation of holes and to define from which diameter (feature) the holes are located. In this case, it would be beneficial to utilize a hole as the tertiary datum, for angular orientation (Fig. 4.6).

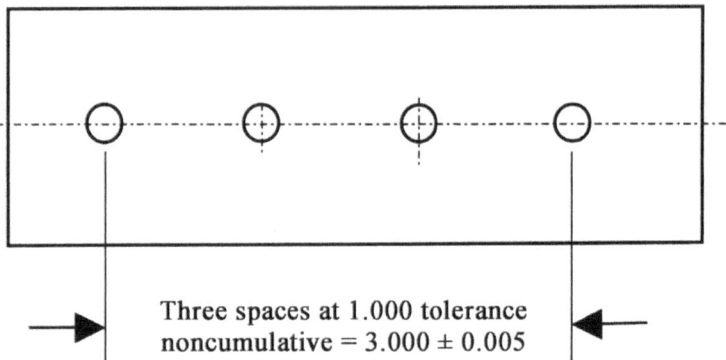

Fig. 4.7 Tolerance noncumulative

Tolerance noncumulative (or nonaccumulative) is an improper method but commonly used to specify hole or feature locations. There is no definition or interpretation of this method in existence. Use positional tolerances in lieu of tolerance noncumulative (Fig. 4.7).

Clearly define from where dimensions are originating. Just a centerline is insufficient unless the feature is clarified by the use of datums. Any dimensions originating from a centerline are originating from an undefined centerline. There many features that could establish a centerline. Use of appropriate datums will define the appropriate datum axis, and a tertiary datum will also provide proper angular orientation where necessary (Fig. 4.8).

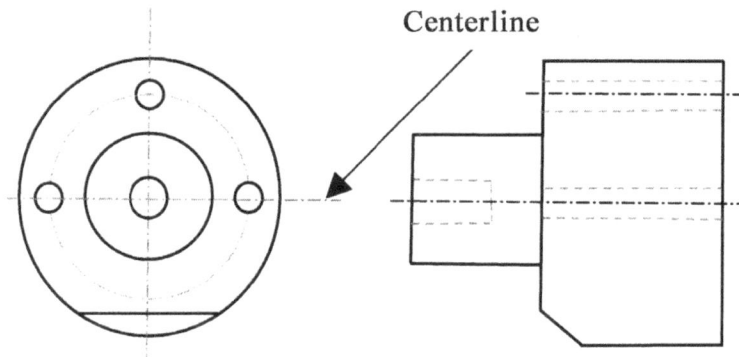

Centerline

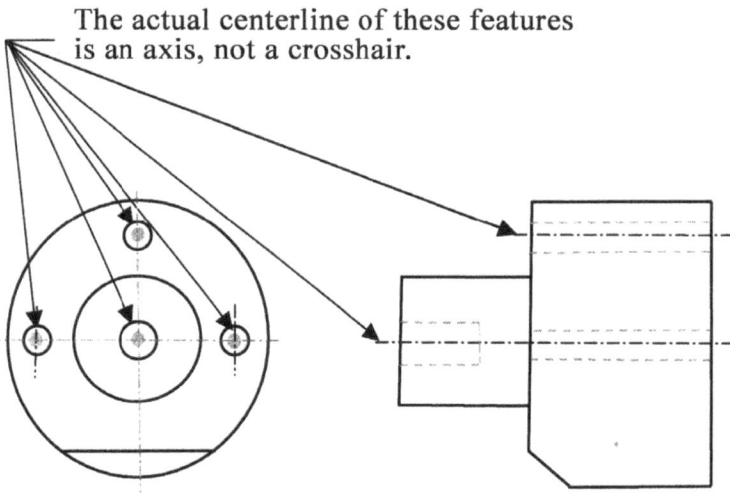

The actual centerline of these features is an axis, not a crosshair.

Fig. 4.8 Centerline and axis

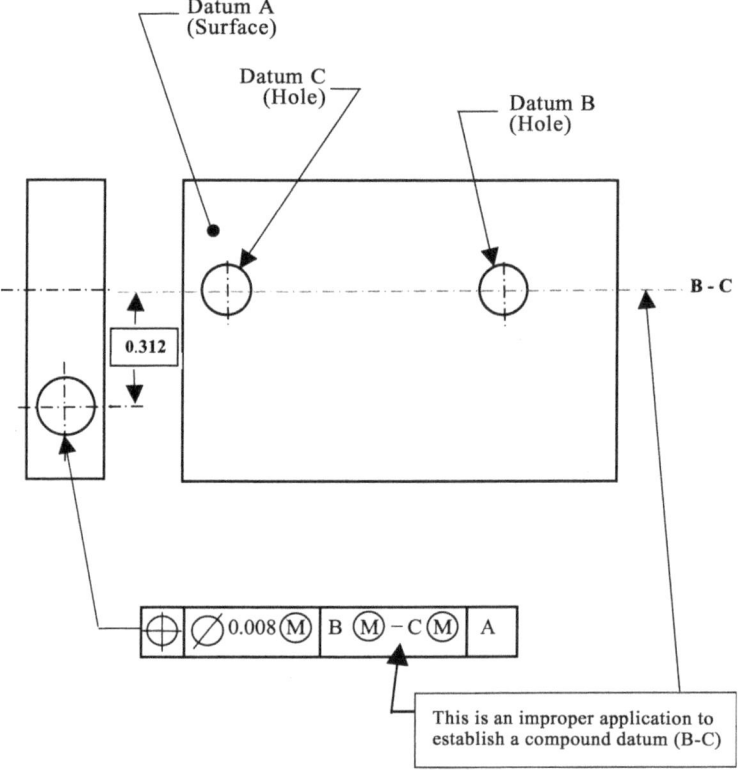

Fig. 4.9 Composite datum

Frequently, an attempt to create a compound datum is performed as shown in Fig. 4.9. Depending on which version of ANSI Y 14.5 is used it may be proper or improper. Regardless of which version is used the main problem is the interpretation of it, especially with MMC, and how will an inspection gage be made. In all probability using Datum B as a primary, and Datum C as a secondary would be sufficient. Datum holes B and C are an axis, that are perpendicular to the surface. The axis runs in a different direction than the length of the part and it is perpendicular to the surface datum A. Also, the use of MMC on the datum holes is very suspect and difficult to determine. Always evaluate how an inspection tool will be made and how tolerances affect your calculations.

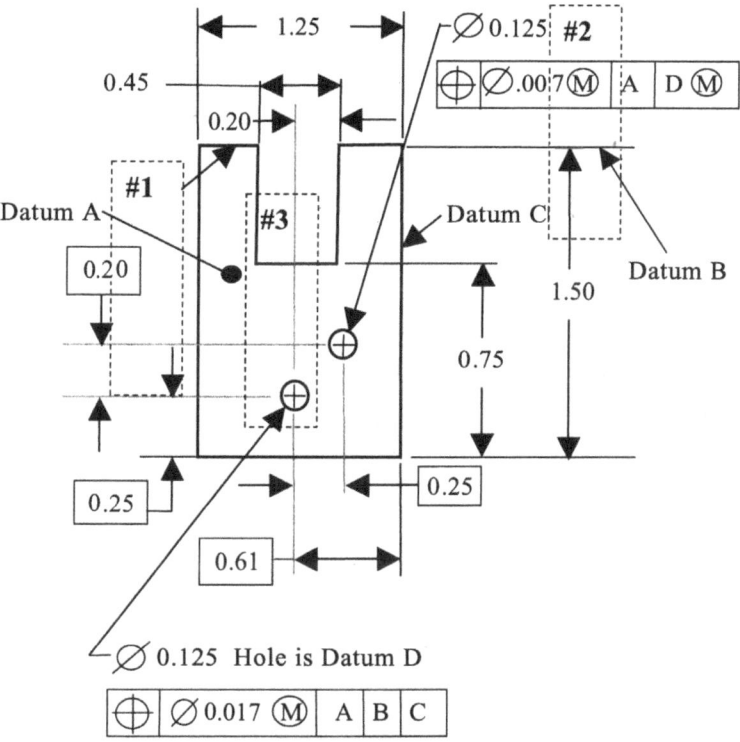

Fig. 4.10 Clarification of intent

Figure 4.10 shows some frequently seen errors: edges not located, slot not located, and angular orientation of hole omitted. The following are areas 1, 2 and 3 in the figure:

#1 One of these surfaces is not located. A coplanarity dimension will locate these two surfaces with respect to each other. Also, it is not defined which surface or surfaces are Datum B. The symbol for a continuous feature will help solve some of the confusion.

#2 This hole is lacking angular orientation. The hole is perpendicular, located from the axis of D, but it has no clocking (angular orientation). A tertiary datum is required.

#3 This interpretation of the location of the slot is very complex because of omitted information. It appears that center of slot is located from either the center of Datum D or from the basic 0.61 dimension. If located from Datum D, then there is no angular orientation and positional tolerance is omitted. Also, if it is located from basic 0.61, GD&T is required. Profile tolerance would clarify the intent.

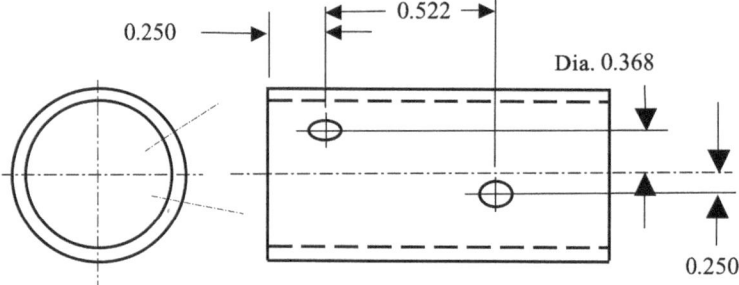

Fig. 4.11 Features origination from a centerline

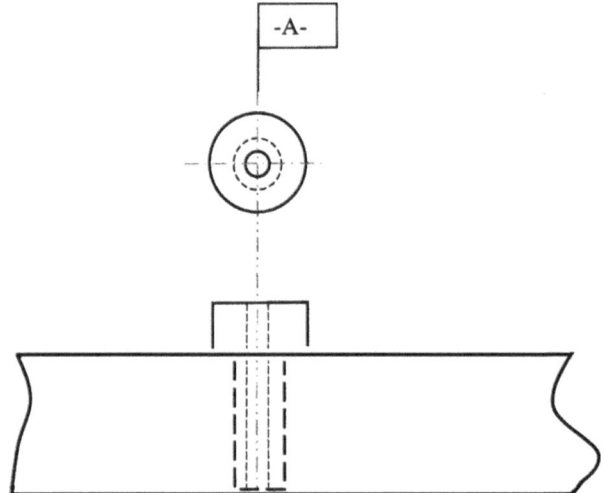

Fig. 4.12 Clarifying datum

A very common example of dimensions originating from a centerline is shown in Fig. 4.11. The centerline from which the holes are being located is undefined and has no meaning. It might be the ID (inside diameter) or the OD (outside diameter). The holes also have no angular orientation with respect to one another.

Frequently, a feature is stated as being located from the center of an object (such as a pin, screw, etc.), or a hole in the part where a pin is installed, is used as a datum. In Fig. 4.12, if it is the center of a pin (or possibly the installation hole) it needs to be defined which feature is the datum. If it is the pin, then the datum needs to be clarified which feature of the pin such as the outside diameter of the head, the center of the installation hole, or the small diameter of the pin is to be used to establish the datum. In all probability, whichever feature is used, RFS should apply. Also, it is important that the datum feature be accessible. If the hole in which the pin is installed or the small diameter of the pin that fits into the hole is the datum, then it is not accessible,

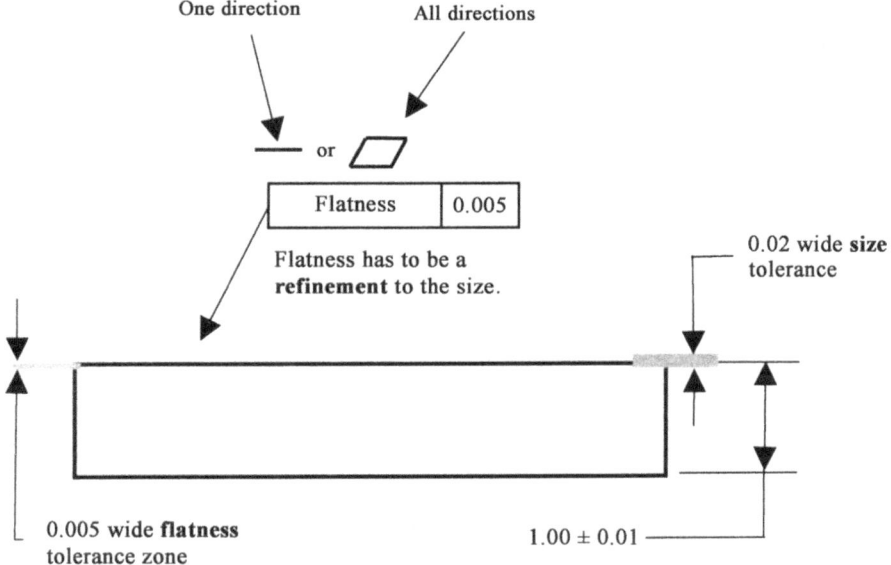

Fig. 4.13 Flatness

Parallelism must be a refinement
to the size tolerance.

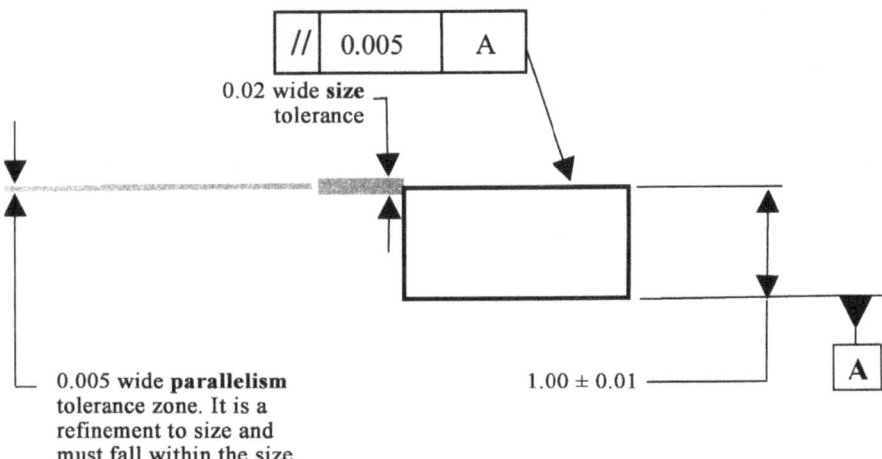

Fig. 4.14 Parallelism

and should not be used. Frequently the designer will justify his selection stating that it is proper as it will be inspected in process, but truthfully he has no control over inspection. The different stages of fabrication, assembly or inspection of the part may also be subcontracted. Also, other users of the drawing may not fabricate in the same stages as the as the original design activity (Figs. 4.13 and 4.14).

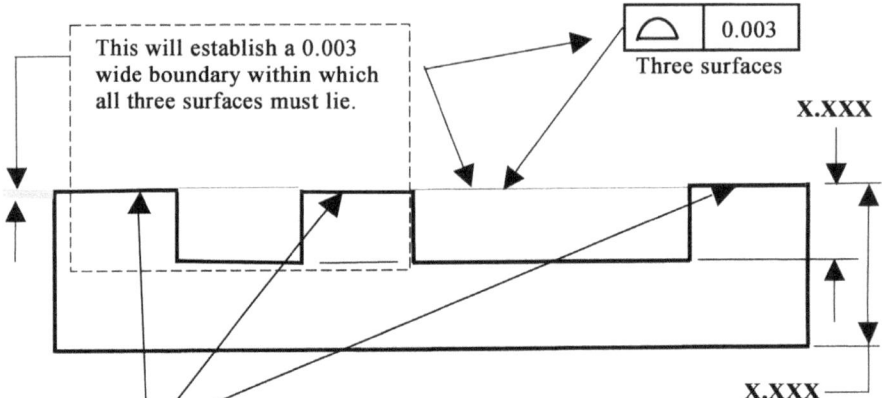

These three surfaces are not located with respect
to one another. Use coplanarity symbol to control
these three features with respect to one another.

Fig. 4.15 Coplanarity

Frequently the relationship of the three surfaces is omitted. The use of coplanarity (same symbol as profile, but no datum references) is the method of specifying
the surfaces relationship to each other. Point to the extension line between surfaces,
use the symbol (same as profile) and the desired tolerances within which the
surfaces have to lay. There is no reference to datums and state the number surfaces.
The symbol for a continuous feature (CF) can also be used and then it will be treated
as if it were one surface (Fig. 4.15).

Chapter 5
Drawing Types and Requirements: Select the Proper Drawing for Your Application

There are many different types of drawings, and each one serves a function in the engineering world. These drawings are used to fabricate parts; inspect, test, assemble, purchase parts; and provide information (such as schematics, installation, interface, interconnect, etc.). Each of these drawing types requires different documentation for the specific item being depicted. The requirement for the fabrication of a casting differs from those requirements for a printed wiring board or a sheet metal part. The documentation requirements of all items originate from ASME Y14.24, ASME Y14.100, and other sub-tier documents that they invoke.

The three most common drawings are *detail drawings, assembly drawings,* and *vendor item drawings*. The most controversial drawing type is the vendor item drawing, as many features are undefined and much of the information is not available.

All the intricate detail requirements of each drawing type will not be completely defined herein, as the requirements are adequately defined within the appropriate governing documents. I will be only stating the appropriate usage of each item, along with some of the basic requirements, and including any pertinent data and the most common mistakes. Always completely review the requirements of ASME Y14.24, ASME Y14.100, and your contract for the type of drawing that you are preparing.

Frequently there are special conditions that occur and preclude complete compliance to the governing contractual documents. Honesty with your customer and the request for a deviation is preferred more than hoping that your variation won't be discovered. It will be! It is just a matter of time. If your deviation is not approved and your deficiency is discovered it may cost time, money and respectability. Parts may be required to be reworked; schedules missed and contractual issues with subcontractors and vendors may ensue.

© Springer International Publishing Switzerland 2015
R. Hanifan, *Perfecting Engineering and Technical Drawing*, SpringerBriefs in
Applied Sciences and Technology 139, DOI 10.1007/978-3-319-06983-8_5

Drawing Notes

The following are a few of the requirements that have been extracted from ASME Y14.100, paragraph 4.26. These are good and logical requirements whether used on Government contracts or commercial programs. Frequently the content of drawing notes is given minimal attention, but their requirements are as important as the delineation of the object and the dimensioning and tolerancing. When I review a drawing I will devote as much time to the understanding of each sentence in a note as I do to the dimensioning and tolerancing of a drawing.

The following are the most common requirements that are neglected and cause interpretation problems.

Notes shall be unambiguous and contain complete acceptance/rejection criteria. They are to be concise statements using the simplest word and phrases for conveying the intended meaning. They shall be grammatically correct statements. When terms "Will", "Should" and "May" are used, the intent of purpose is nonmandatory and it is not inspectable. This type of information will just cloud other pertinent data in the notes.

The *arrangement of notes shall not be* interpreted as an order of precedence or sequence in manufacturing or assembly unless so specified.

Indefinite terms   Never use terms such as and/or, etc., e.g., and i.e.

Reference to other documents for the purpose of specifying requirements or drawing interpretation shall be as specific as possible. The whole of a given document provisions shall not be made applicable by reference unless all of its provisions are required. Reference to *paragraph numbers in other documents shall not be made*. Usually a reference of the subject matters pertinent subject is sufficient.

Example: "Drawing notes shall be in accordance with requirements of ASME Y14.100."

DO NOT make reference by paragraph number: EXAMPLE: "Drawing notes shall be in accordance with paragraph 4.26 of ASME Y14.100.

If *all* the requirements of ASME Y14.100 are to be invoked then the note would be written as: "Interpret drawing in accordance with the requirements of ASME Y14.100."

Reference shall be made to a method, identified requirement, class, grade, or type. Always verify that all the data is specified in the note or list information such as "Class optional" when that is desired.

Reference shall be made only to documents whose technical accuracy and *availability are assured*.

When a portion of a document is applicable, the extent of its applicability shall be stated. However, *reference to paragraph numbers in other documents shall not be made*.

Do not include in notes contractual requirements, such as statements of costs, time and place of delivery, methods of payment and requirements for submission, approval, or distribution of data, reports, or plans.

Reference to standardization documents shall be by basic identifier, *excluding revision level*, except when identification of a specific issue is essential to drawing interpretation.

Nonstandard symbols or annotations other than flag-notes shall be defined in the notes. Include any non-standard abbreviations that are used on the drawing.

Note numbers of deleted notes shall not be reused. When notes are deleted the common practice is to leave the number and the contents blank although I have seen where some organizations enter the word "deleted" however it is not necessary.

Drawing Note Content

The drawing notes provide pertinent data in word form to complement the delineation or to clarify the requirements of the parameters (acceptance or rejection criteria). They do not contain ambiguous, uninspectable parameters. Controls shall be established for all variables that could affect the accuracy of test results. All parameters are to contain a tolerance or a range of values (directly or indirectly) that are inspectable.

Notes shall be short, concise statements, using the simplest words and phrases for conveying the intended meaning. The use of abbreviations should be minimized and confined only to recognized and authorized abbreviations. Sentences should be written in the affirmative manner where the requirement is mandatory. Requirements shall be specified as requirements, not methods or actions. For example,

Proper: Part shall be free of burrs and sharp edges.
Improper: Remove all burrs and sharp edges. (This is an action, not a requirement and assumes that the part contains burrs and sharp edges.)

Commonly, the drawing notes supplement the drawing by providing the following information. This is a list of the types of items specified in the notes. It is not all inclusive; it provides just the most common information.

- Material
- Finish
- Supplemental dimensional data

    Surface texture
    Fillet and corner radii
    Sharp edge and burr requirements
    Marking requirements

Mandatory manufacturing information

Material processing requirements such as stress relieve

Cautionary information

Applicable geometric dimensioning and tolerancing (applicable specification and date) applicability

Installation information for rivets, inserts, pins, etc.

See ASME Y14.24 and ASME Y14.100 for complete requirements.

Drawing Titles

Although this is not a major item, it is an item that is very visible to all users of a drawing. Drawing titles are a very simple item but many organizations make it complex by ignoring the rules. Do not confuse a nomenclature and type designation with a drawing title. Nomenclatures and type designations are assigned by customers in accordance with Mil-Std-196. The rules for drawing titles are logical and very few. Drawing titles are governed by ASME Y14.100. It is the name by which the drawing or item will be known and consists of a basic name and modifier, when required, to differentiate like items.

The following are most of the rules that apply to drawing titles.

(a) The title shall be *as brief as possible*, describe the item and distinguish between similar items.

(b) The title shall consist of a noun or noun phrase (basic name). Modifiers may be used to distinguish between items with the same basic name.

　(1) A modifier may be a single word or phrase. The first modifier narrows the concept established by the basic name, and succeeding modifiers continue the process.

　(2) The conjunction "OR" and preposition "FOR" shall *not be used*.

(c) The noun or noun phrase establishes the basic concept of an item.

　(1) A compound noun or noun phrase is used when a single noun is not adequate.

　(2) The noun or noun phrase describes the part and the usage of the part, *not material or method of fabrication.*

(d) The noun or noun *phrase shall be used in singular form*, except:

　(1) Where the only form of the noun is plural, such as in "TONGS".

　(2) Where the nature of the item requires the plural form, such as "GLOVES".

　(3) Where multiple single items appear on the same drawing such as "FUSES".

(e) Abbreviations *should be avoided*.

(f) The title shall be consistent with the title of the next assembly.

(g) Reference to major assemblies or end items *shall not be used* except when necessary to differentiate similar items.

(h) Non-part drawings, such as schematic diagrams, shall include the drawing type as part of the title. For example:

TRANSFORMER ASSY, SCHEMATIC DIAGRAM

(i) An ambiguous noun is not used alone but may be used as part of a noun phrase. For example:

Preferred	Not Preferred
CIRCUIT CARD ASSEMBLY	ASSEMBLY, CIRCUIT CARD
PRINTED CIRCUIT BOARD	BOARD, PRINTED CIRCUIT

The most common mistakes that are made are the inclusion of Reference Designations and program names in the title. These should be avoided and they also may preclude multiple usages.

Many times the preparer of the drawing has no control over the drawing title as it may be established and controlled by other individuals or groups. When a drawing title error is discovered it is recommended that it be tactfully brought to the attention of the individual or group responsible of the infraction.

Drawing Statements

All Drawings (When Specified in the Contract)

There are many types of drawing statements that may be required on your drawings and parts list. Verify your company requirements and determine whether there are any contractual requirements. Some of the many drawing statements and/or identifications that may be required are as follows:

- Company Proprietary Statements
- Limited Rights Legends
- Distribution Statements
- Export Control Notices
- Destruction Notices
- Government Purpose Rights
- Critical Safety Items
- Electrostatic Sensitive Devices
- Explosive Items

These are seemingly meaningless items and of not much significance, however, when omitted by you or your subcontractors, it can cost enormous amounts of money to correct their omission. I have seen expenditures of millions of dollars to fix a simple omission of these statements.

Additionally, always verify the exact wording of these required statements by reviewing your engineering procedures and/or contractual requirements.

Mono-Detail or Multi-Detail Drawing

The delineation of parts may be accomplished by many different methods. Each drawing type is unique and contains its own particular requirements, and when the drawing is prepared as a multi-detail drawing, it must be permitted by the contract.

Occasionally, drawing types or categories are combined on the same drawing. While this may be permissible (verify with your contract), the decision to combine drawing types should be made cautiously. Significant benefits should outweigh such potential disadvantages as:

- Increasing the complexity, which may diminish the clarity and usefulness.
- Accelerated change activity of the combined drawing, which may increase the need to update associated records, material control data, manufacturing planning, and so on.

Mono-Detail Drawings

The use of mono-detail drawings will be specified on the contract or in your engineering procedures. A mono-detail drawing depicts a single part per drawing. Neither the depiction of details on assemblies and inseparable assemblies nor tabulation of parts (detail parts or assemblies) is permitted. A mono-detail detail drawing shall comply with the applicable requirements of its drawing category.

When mono-detail drawings have been selected by the contract, it is recommended that the contract be tailored to allow for certain conditions such as inseparable assemblies, various configurations of assemblies, and tabulated drawings.

A drawing depicting "*shown*" and "*opposite*" parts using a single set of views is considered to be a tabulated mono-detail drawing.

Tabulation of Drawings

All Drawings may be tabulated to delineate similar items which as a group, have some common characteristics and some variable features.

Multi-Detail Drawings

The use of multi-detail drawing must be permitted by the contract or your engineering procedures. A multi-detail drawing depicts two or more uniquely identified (each part shall have a part identifying number) parts in separate views or in separate sets of views on the same drawing. It is prepared for parts that are usually related to one another. The use of multi-detail drawings should be made cautiously. The same revision status applies to all details on a multi-detail drawing; therefore, a change to one detail of the drawing may affect the associated records of all other details (material control data, manufacturing planning, and so forth). Significant benefits should outweigh this potential disadvantage, as well as such others as diminished clarity and usefulness resulting from increased drawing complexity.

See ASME Y14.24 for complete requirements.

Detail Drawing

Part Fabrication

This drawing *completely defines the complete end product* requirements of a part. It defines configuration, dimensions, tolerances, materials, finishes, marking, surface texture (if necessary), and any mandatory processes. It is important to ensure that all tolerances are applied correctly and that all data contains complete acceptance/rejection criteria. Item Identification (Part identification number) shall be established.

If any dimensions originate/terminate from a centerline, ensure that they are clearly identified with the use of geometric dimensioning and tolerancing. All dimensional requirements are to be specified such that there can be only one interpretation. Review notes to ascertain that all requirements contain acceptance/rejection criteria and that their intent in unambiguous.

See ASME Y14.24 for complete requirements.

Matched Set Drawing

A matched set drawing is a special application drawing that delineates items that are matched and for which replacement as a matched set is essential. Matched parts are those parts, such as special application parts, that are mechanically or electrically matched, or otherwise mated, and for which replacement as a matched set or pair is imperative. A common depiction are parts that are precision machined with other parts and replacement of the parts would not be functional if delivered individually and not as a set. It is mandatory that spare parts be delivered only as matched sets.

A matched set drawing is prepared when the required dimensions, tolerances, or other characteristics of items can only be specified in terms of the matched relationship. This includes items that are interchangeable only as a set because of special requirements for machining, electrical characteristics, performance, and so on. Under such conditions, a matched set drawing defines the matching relationship. Individual parts of the set may be delineated by the matched set drawing or by other drawings.

The matched set drawing shall include, as applicable:

a. The physical or functional mating characteristics of the matched items (set).
b. A unique identifier (part identifying number) assigned to each of the parts and to the matched set.

   Part Marking: In addition to the part identification marking of the set, it is recommended that the matched parts be marked with the word "set" next to the part identifying number. Previously it was mandatory to mark these parts with the word set, but it no longer is a requirement.
c. The statement "furnish only as a matched set" or similar note shall be included on the drawing.
d. The drawing shall comply with all the other requirements for the drawing category.

See ASME Y14.24 for complete requirements.

Assembly Drawing

Assembled Parts

There are two different types in this category: a separable assembly and an inseparable assembly. See ASME Y14.24 for complete requirements. A detail part that contains threaded inserts, riveted parts, and welded parts/members is normally construed as an inseparable assembly.

The intent of an assembly drawing is to define the item as an end product. The data shall be presented in such a manner that the requirements are not subject to more than one interpretation. Assembly drawings should not be structured to mimic assembly or inspection sequences. Drawings should *depict logical levels of assembly or disassembly*, *at testable levels*, as a functional item or a *deliverable item*.

Assembly drawings should not contain noncritical manufacturing processes or be structured to mimic assembly or inspection sequences, as previously discussed, as increased costs may be realized such as the following:

a. The total quantity of drawings requiring preparation, maintenance, and delivery to the customer will increase.
b. Maintenance costs of the drawing increase as a result of additional engineering changes being written against the documentation so as to revise processing data

or manufacturing information as techniques change or as changes to equipment availability occur. When the drawings are structured logically and manufacturing process information is omitted, then manufacturing may revise its own processes without the costly burden of revising the drawing on each occurrence.

c. *Spares levels*—When assemblies are not created logically, additional maintenance, repair manuals, and testing data are required to support these extra intermediate levels.

d. *Stocking levels*—Intermediate assembly levels will be required at each of these levels to support the sparing requirement, which creates additional administrative and manpower burdens.

e. *Testing levels.* Tests and equipment are required to support each level, along with appropriate documentation. Test requirement documents, specifications, and technical manuals may be required to support these additional levels.

f. *Reference designations.* Assignment of unnecessary and/or improper reference designations complicates the identification of items and their locations. These designations may become so large that they become illogical and difficult to mark on the equipment and list in technical manuals as well as system and schematic diagrams. See Chap. 6.

g. *Technical Manuals and Instructions.* Technical manuals and other instructional manuals will become more complex than necessary when assemblies are not created at functional and logical levels.

Attaching Parts

Attaching parts (bolts, nuts, washers, and so on) that are required to mount assemblies into their next higher assemblies or on foundations, shall be called out *on the parts list of the drawing that defines the attachment* (usually the next level assembly or installation drawing).

Reference Items

A cross reference to applicable installation drawings, wiring lists, schematic diagrams, test specifications, and associated lists shall be provided as applicable.

Assemblies Containing Electrical Parts

Testing data in the form of a standalone test specification or test requirement is normally required on assemblies where *active circuitry* is employed. Testing data is of extreme importance to customers and will become an area of intense inquiry.

Caution should be exercised to restrict testing to actual testing requirements and not to use electrical inspection requirements as testing data. A common occurrence is an "electrical continuity test" to determine if all electrical connections have been made on wiring or circuitry. The actual requirement on the drawing is to make all the connections listed properly. An electrical continuity test is solely an inspection to verify that the assembly has been properly fabricated, it is not a test.

Location of Parts of the Assembly

Locate all parts and materials and any special requirements that are performed or installed at this level.

Electrical Parts of the Assembly

Identify, in the field of the drawing, the reference designation assigned to the respective parts in the parts list (BOM). See Chap. 6 for reference designations.

Quantity of Parts

Quantities in the Parts List (BOM) shall agree with the field of the drawing and or notes. The definition and usage of most bulk items will appear in the notes.
    See ASME Y14.24 for complete requirements.

Inseparable Assembly

The same requirements apply as an assembly drawing except as follows.
    An inseparable assembly, although documented as an assembly drawing, actually is a part and is identified as a part. The inseparable assembly drawing delineates items (pieces) that are separately fabricated and are permanently jointed together (as in welded, brazed, riveted, sewed, glued, or otherwise processed) to form an integral unit (part) not normally capable of being disassembled for replacement or repair of the individual pieces.
    An inseparable assembly drawing may be prepared in lieu of individual mono-detail drawings for inseparable parts, but it must be allowed by the contract, either by the allowance of multi-detail drawings or tailored when mono-detail drawings have been selected as an option in the contract. Always verify within the contract that it has been tailored to allow multi-detail drawings for inseparable assemblies.

The drawing shall disclose all the required data of an assembled, end product item, complying with all the applicable requirements of a detail drawing and assembly drawing (including a parts list or equivalent disclosure of individual parts). Materials, finishes, envelope configuration, complete dimensional characteristics, and so forth shall be provided.

As the item is being defined as an end item product, frequently many of the individual features may be omitted and left to the ingenuity of the fabricator (such as end joint features of a welded or brazed assembly), with only the end item product being defined. The detailing of the individual pieces on separate drawings is not normally a very cost effective method.

See ASME Y14.24 for complete requirements.

Detail Assembly

A detail assembly drawing depicts an assembly on which one or more parts are detailed in the assembly view or on separate detail views. Even though this is a widely used drawing, it is a drawing category that no longer is recognized in ASME Y14.24 and falls into the category of a multi-detail drawing. The preparation of a detail assembly drawing in lieu of individual mono-detail drawing must be allowed by the contract, either by the allowance of multi-detail drawings or tailored when mono-detail drawings has been selected as an option in the contract. It must be tailored to permit the use of multi-detail drawing for a detail assembly.

The detail assembly drawing shall provide complete end product definition of each individual part and the assembly, as each part may be spared as a replaceable item.

The drawing shall contain all the end item disclosure information of an assembly drawing and detail drawing. Parts shall contain complete definition of configuration, dimensions, tolerances, materials, finishes, marking (if necessary), surface texture (if necessary), and any mandatory processes. Ensure that all tolerances are applied correctly and that all data contains complete acceptance/rejection criteria.

Interface Drawing

This drawing depicts physical and functional interfaces of related or co-functioning items. It does not establish item identification (no part number). It is used to establish and maintain compatibility between items having a common boundary. It coordinates and controls interfaces between related or co-functioning systems and communicates design decisions to participating design activities. This drawing does not normally belong as a part of your Technical Data Package as it fabricates no parts and it is advisable to not to have them included as part of your Parts List or drawing notes. When this is necessary because of your activities processes, it is recommended that they be shown as "Reference Items" and not contractually binding on other users of your drawings.

This drawing is used to control mechanical, electrical, hydraulic, pneumatic, interconnections, configuration, installation, operational sequence requirements, and system switching interfaces.

These drawings are normally used to convey design information to other activities, but I have seen where they are used to convey requirements to other departments within the same design activity. This practice gives the appearance of a lack of communications and cooperation with other departments and can result in an adverse appearance of your company.

See ASME Y14.24 for complete requirements.

Altered Item Drawing to Alter an Existing Item

This drawing physically alters an existing item (a usable item) *under the control of another design activity*. You cannot alter your own items or items under your design control. If this were permitted, then you would build a part incorrectly and then perform the modification.

The item's form, fit, function, and performance requirements prior to the alteration must be described in some manner, and you must provide a complete definition of the alteration. This can be accomplished in many ways. As the item existed prior to the alteration, it may be described on the actual altered item drawing, on a separate drawing, or by a vendor part number.

- Always use caution when altering a vendor supplied part; by virtue of this alteration, any warranty could be compromised.
- The original part number shall be provided.
- The original part number should be removed or obliterated and the part reidentified with the new part identifying number. Microcircuit re-identification marking shall be in addition to the existing original marking and shall be visibly separate from and in no way interfere with the existing marking.
- The alteration may be performed by you or by another competent manufacturer, including the original manufacturer or a third party.

An Altered Item may be delineated on the using assembly drawing provided the prerequisites of the section "REQUIREMENTS" of an Altered Item of ASME Y14.24 are met.

See ASME Y14.24 for complete requirements.

Printed Board and Discrete Wiring Board Drawing Sets

A printed board and discrete wiring board drawing sets consist of those drawing which define the configuration of printed wiring or printed circuit and discrete wiring boards and assemblies. They establish requirements for board fabrication, board assembly (mounting of electrical, attaching hardware, etc. and test. The term board includes all types of dielectric base materials (flex, rigid or a combination thereof). The term "printed" applies to all circuitry patterns which are etched, deposited, screened, or bonded to the base material. The term "discrete wiring" applies to circuitry patterns that are defined by the routing and terminating of discrete wires to form point-to-point electrical connections."

A typical drawing set may include, as applicable:
Assembly drawing.
Schematic diagram.
Master drawing (Printed Wiring Board Detail Drawing).
Master pattern drawing.
Artwork.
Artwork master.
Production master.

The assembly and master drawing establish item identification for the assembly and board.

Printed board drawing sets are prepared to describe and control printed wiring or printed circuits boards and assemblies.

Discrete wiring board drawing sets are prepared to describe and control discrete wiring boards and assemblies. Boards may be single-sided, double-sided, or multilayer.

An assembly and master drawing are normally required to provide complete end item definition, but the use of a master pattern drawing, artwork and artwork master, and/or a production master is dependent on the type of board being designed, how it will be fabricated, and how the documentation will be used.

Master Drawing: Detail Drawing

In some manner a definition of the circuitry is required. Circuitry definition may be in the form of a software drawing (Digital Form of Printed Boards), a separate master pattern drawing, or inclusion of a representation of the circuitry, from the printed wiring master drawing, as subsequent sheets of the drawing. The depicted circuitry is the circuitry prior to etching and not that of the actual finished printed wiring board. The preferred method is to furnish software (digital data) that defines the circuitry. The use of a master pattern drawing and definition of the circuitry as subsequent sheets is not preferred. Storage problems and usability are the main issues with anything other than a software definition.

The end item requirements, materials, and physical shape of the board require definition. This is normally straightforward. The board outline is normally defined using basic dimensions and a profile tolerance, originating from a surface and two datum holes (indexing holes or tooling holes). Normally a board edge should never be used as a datum; however an edge may be used when it represents a major function of the printed board. Normally the board edge is actually located from the primary datum hole, and the symbol of a "dimension of origin" is frequently used for this clarification.

PRINTED BOARD MASTER DRAWING—A document that shows the dimensional limits or grid locations that are applicable to any and all parts of a product (printed board) to be fabricated including the arrangement of conductors of nonconductive patterns or elements, the size, type, and locations of holes, and all other necessary information. Reference ANSI/IPC-D-325 and IPC 2221 as applicable for drawing requirements.

DISCRETE WIRING BOARD MASTER DRAWING—A document that includes all information to guarantee fit, form, and function of the discrete wiring board and shall establish as a minimum: interconnection techniques, wire type and gauge, size, shape, location, and tolerance of all discrete wiring board features. Material, plating, marking and coating requirements shall also be indicated as applicable. Reference ANSI/IPC-D-325 and IPC 2221 as applicable for drawing requirements.

ARTWORK—An accurately scaled configuration that is used to produce the artwork master, production master, master pattern drawing. Reference ANSI/IPC-D-275, ANSI/IPC-D-325 and IPC 2221 as applicable for documentation requirements.

ARTWORK MASTER (*Photoplot Master*)—An accurately scaled usually 1:1 pattern that is used to produce the production master. Reference ANSI/IPC-D-275, IPC-2221 and ANSI/IPC-D-325 as applicable for documentation requirements.

PRODUCTION MASTER—A 1:1 scale pattern that is used to produce rigid or flexible printed boards within the accuracy specified on the master drawing. Reference ANSI/IPC-D-275, IPC-2221 and ANSI/IPC-D-325 as applicable for documentation requirements.

MASTER PATTERN DRAWING—The master pattern *drawing* is a reproduction of the original artwork, artwork master, or data base prepared on drawing format. Reference ANSI/IPC-D-275 and IPC-2221 as applicable for drawing requirements.

DIGITAL FORM OF PRINTED BOARDS—Automated layout techniques (computer-aided) may be used to generate various aspects or descriptions (drill data, artwork, testing, etc.) of the printed board information and to document a digital form of the drawing that are a part of the printed board drawing. Reference ANSIIPC-D-350 for the requirements on generating printed board descriptions in digital form and ANSI/IPC-D-351 for generating a digital form of printed board drawings.

Holes Sizes and Locations

Commonly hole locations are omitted by accident. Holes shall be located dimensionally or by annular ring requirements. Holes that have no conductive material (unsupported holes) and holes that are in large circuit patterns/pads require dimensional locations, as the annular ring requirement is not sufficient. It may be unrealistic to define the size and location of small via holes (such as those used in printed wiring boards utilizing surface mount components or interconnects) as an inspection requirement, as they will probably be solder filled. It is recommended that these via holes be defined only by reference dimensions or allowed to be located by annular ring. It is further recommended that hole quantities be omitted if a hole chart is utilized. Normally hole sizes are indicated by a coding method and each hole is indicated on the face of the drawing.

Board Assembly Drawing

PRINTED BOARD ASSEMBLY DRAWING—The document that shows a printed board, separately manufactured components, and any information necessary to describe the joining of them together in order to perform a specific function. Reference ANSI/IPC-D-325 and IPC 2221as applicable for drawing requirements.

DISCRETE WIRING BOARD ASSEMBLY DRAWING—A document that shows a discrete wiring board separately manufactured components, and any information necessary to describe the joining of them together in order to perform a specific function. Reference ANSI/IPC-DW-425 for drawing requirements.

Terms

Confusion also lies with many of the terms used in printed wiring board documentation. The following are definitions of some of these key terms that may affect documentation.

Annular Ring That portion of conductive material completely surrounding a hole.

Artwork Master, Printed Wiring Master Pattern Drawing (Stable Base Artwork) The master pattern drawing is a reproduction of the original artwork or database prepared on a drawing format.

Discrete Wiring applies to circuitry patterns that are defined by the routing and terminating of discrete wires to form point-to-point electrical connections.

Printed Wiring Drawing This was previously known as a *printed wiring master drawing* or *master drawing*. It is the drawing that shows the dimensional limits or grid locations applicable to any or all parts of a printed board (rigid or flexible), including the arrangement of conductive and nonconductive patterns or elements; size, type, and location of holes; and any other information necessary to describe the product to be fabricated.

Production Master A one-to-one scale pattern used to produce one or more printed boards within the accuracy specified.

Via Hole A plated through hole used as a through connection, but for which there is no intention to insert a component lead or other reinforcing material.

See ASME Y14.24 and sub-tier ANSI/IPC standards such as ANSI/IPC-D-275 (IPC-2221), ANSI/ IPC-D-275, ANSI/IPC-D-350, ANSI/IPC-DW-425, IPC-2222, IPC-T-50, and ANSI/IPC-D-351 for a complete definition of all requirements.

Control Drawings

These types of drawing are used when you want to procure a part from a subcontractor or a vendor. When you want to purchase a vendor part, the documentation will appear on one of the following drawing types.

Procurement Control Drawing
Vendor Item Drawing
Envelope Drawing
Source Control Drawing

The most common drawing is a Vendor Item Drawing. This drawing buys a part off the shelf with no restrictions other than what the vendor normally provides.

An alternative to preparing a drawing is to list the vendor and his part number on your parts list. Problems arise when only a vendor part number is used. This restricts you to buying only that part and only that vendor is allowed, unless the drawing is revised. Competitive procurement is eliminated. Additionally, verify that you are contractually permitted to list a vendor part number in lieu of preparing a drawing.

A control drawing normally does not cost much to prepare and, when done properly, is much preferred to listing only a vendor part number on a parts list.

Vendor Item Drawing

To Procure a Part Right Off the Shelf

If you have a Government data requirement that requires that you prepare vendor item drawings, this will be the most controversial drawing in your technical data package, because it does not provide a complete item description. You are only defining what the vendor guarantees and within that, the features that you consider critical. This drawing provides an engineering description and acceptance criteria for commercial items or vendor developed items that are procurable from a specialized segment of industry.

When you want to buy a part from a vendor and it comes right off the shelf, then you prepare a vendor item drawing. This is just like buying it from a store. It is just as they advertise it, with no modifications, selections, or restrictions that they do not advertise or guarantee. This drawing allows procurement to have competitive bids from multiple vendors. Purchasing can buy from anyone that meets the qualifications you have listed on the drawing. List the vendor (two or more if possible), the physical definition, and any other pertinent information from the vendor. Never tighten tolerances beyond what the vendor guarantees or relax any noncritical dimensions and characteristics. The most important item on the drawing is the name of the vendor and his part number. Regardless of what other information is on the drawing, the part number is the item that is procured. The information in the notes and the outline is for inspection purposes only; to verify that the part purchased complies with the information depicted on the drawing.

In lieu of a drawing, you could list a vendor part number in the BOM/Parts List, but that restricts purchasing to buying only that part from the vendor listed. This action can be taken only when contractually permitted.

Do not list recommend mounting holes on the drawing. Many times, the vendor's data sheets will contain errors, and there can also be slight differences between vendors, depending on who supplies the part.

The part identifying number is the vendor's part number, not your part number (although, in many cases, a vendor will use your part number as his part identification). You may desire to have your part identification, in addition to the vendor part number, marked for administrative purposes. The method used to mark the part is the vendors method, not yours unless they are one and the same. Do not place requirements in addition to those normally provided by the vendor.

When restrictions are placed on a part beyond what the vendor guarantees, then it will no longer comply with the parameters defined by his part number. Relax tolerances and requirements that are not critical to your application, but completely define all critical requirements, provided that they are guaranteed by the vendor.

See ASME Y14.24 for complete requirements.

Procurement Control Drawing

To Buy a Part in Development and Establish a Part Number

This is something of a mix between a vendor item drawing and an envelope drawing, except that it can establish a part number. It is used for vendor or subcontractor items under development when engineering needs a part number to list in their system. A procurement control drawing may be prepared in lieu of other types control drawings to specify criteria for:

- A purchased item
- The alteration of a purchased item or an item defined by a nationally recognized standard
- The selection of a purchased item or an item defined by a nationally recognized standard
- The development and qualification of a new item
- Item identification

Data should be provided to properly define the item, including performance, envelope dimensions, procurement information, and qualification and acceptance requirements. The drawing should contain all the necessary information to define the requirements of your application.

See ASME Y14.24 for complete definition of requirements.

Envelope Drawing

For Items Subcontracted for Development

When a part is subcontracted to a vendor or manufacturer for development, necessary information (such as necessary performance requirements, interfaces, envelope size, and any other pertinent data) shall be shown on the drawing. This drawing shall provide the basic technical data and performance requirements necessary for development or design selection of an item.

No part number is established at this time. If you attempt to establish a part number prior to development being completed, it will not be known whether to use a detail part number, assembly, vendor, or source control number.

When development is completed, the drawing shall evolve into formal documentation, such as a control drawing of some type, a detail drawing, or an assembly drawing.

See ASME Y14.24 for complete requirements.

Source Control Drawing

To Restrict the Purchase of a Vendor's Item

This drawing has a once-in-a-million application, so it will not be discussed extensively. Commonly it may be required because of a customer's contractual request. This type of drawing defines a unique item that requires prior qualification in advance of procurement and restricts procurement to only the vendor listed. It provides a means of establishing engineering requirements for the selection, qualification, and acquisition of an item from commercial sources. It also provides identification of the items/sources qualified to meet the stated requirements for the specific critical application.

The drawing shall disclose documentation to assure interchangeability of specified items, in the stated application, each time acquired.

It provides an engineering description, qualification and approval requirements, and acceptance criteria for commercial items or vendor developed items procurable from a specialized segment of industry. Note, if qualification is not required, the item is normally a candidate for a Vendor Item control drawing. It provides performance, installation, interchangeability, or other characteristics required for critical applications.

Note: "Critical application" refers to those uses where failure of the item would result in one or more of the following conditions:

(a) Endanger life or cause personal injury.
(b) Loss or damage to equipment.
(c) Degrade the performance to a point that would jeopardize its capacity to fulfill its intended function.

Never prepare a source control drawing and list yourself as a sole source of supply for the item.

The requirements are identical to those of a vendor item except for identification of critical application(s), qualification and approval requirements, item identification, approved sources of supply in lieu of suggested sources of supply, and required notation; however, there may be unique Government requirements specified in the contract.

See ASME Y14.100, ASME Y14.24, and your contract to ascertain if there are any special requirements that apply.

Associated Lists

Parts List

Parts lists (PL) are usually not a problem, and whatever format you use is normally acceptable. It has been time proven and it works, or it wouldn't be there. If prepared for a Government contract, consult ASME Y 14.34 to ensure compliance. In areas of noncompliance, it is easier to request a waiver than to change your entire system for a single program.

If, on your contract, you are required to prepare data lists and/or index lists, review ASME Y14.34 to verify your ability to prepare them properly. They are very tedious and complicated documents to prepare properly. Most people will never come into contact with a data list or index list; these documents are used primarily by your customer.

Data List

A data list (DL) is a very complex item. It will list all engineering drawings, associated lists, specifications, standards, and subordinate data lists pertaining to the item to which the data list applies, along with essential in house documents necessary to meet the design disclosure requirements except for in-house documents that are referenced parenthetically. In actuality, a data list is a complete listing of all information appearing on your drawing. The required data includes *document number, revision status,* and *nomenclature.* Optional information is *drawing size* and *number of sheets.*

Index List

An index list (IL) is not normally required in a contract. However, when required, it is a tabulation of data lists and subordinate index lists pertaining to the item to which the index list applies.

For all the above drawing types consult ASME Y14.34, ASME Y14.100 and your contract.

Isometric Views (Value Questionable)

Isometric views can be valuable although they are not really required. They can add some noteworthy information to the drawing on complex parts. On simple parts they are not of much value and only tend to add clutter and take up space and become a part of the drawing review. They are relatively cost free, but if not maintained, they may become incorrect as a result of drawing changes. Since almost all drawings are CAD prepared the value is even more questionable. When models are available to other users, then the value is even less valuable. When no value is added to the drawing I recommend that these views be omitted, especially since there is no requirement for the drawing to contain this data.

Datum Location

The primary advantage is that this isometric view can be used to display a quick look at the *location of datums*. This provides quick orientation as to their location, as opposed to sorting through many sheets or views to locate datums.

Occasionally, other bits of key information such as markings, parting lines, unique finishes requirements, location of parts of an assembly, and so forth, can be displayed on this view. The view should be labeled as *"Reference View"* only, and any of the data displayed on the isometric view shall be shown on the appropriate views of the drawing.

Frequently, the isometric view is utilized in the preparation of technical manuals or other similar information. On an assembly drawing, they provide excellent information that can also be exported into technical manuals. Frequently, exploded isometric views are used in lieu of the traditional orthographic views. Verify that there is no conflict, contractually or internally, before using these views.

Chapter 6
Electrical Reference Designations

If your drawings, procedures or contract invoke ASME Y14.100 then electrical reference designations are required and shall be assigned as defined in IEEE Std 315 and ASME Y14.44.

Electrical items are assigned an identifying letter and number for quick identification and location in equipment. It is called a Reference Designation and it is letters or numbers, or both, are used to identify and locate discrete units, portions thereof, and basic parts of a specific set. (Do not confuse a reference designation with a function designation of an item. See the following definition).

Each type of electrical item has a reference assigned for quick identification. For example, a resistor is assigned the letter "R." Then each item is assigned a number identifying it uniquely. When this is all put together, it specifies exactly where an item is located in the equipment. It is really nothing more than an address with an identifying letter of the occupant. Take, for example, a resistor R1. When the complete prefix is used, "1A2A3A1A1*R1*" it will quickly reveal exactly where this particular component is located. Think of this as your address:

R1	This is you
A1	558
A1	Main Street
A3	Chicago
A2	IL
1 (unit number)	U.S.A.

The marking of reference designations is extremely beneficial for rework and troubleshooting of equipment. Normally, only a partial reference designation is marked, and it is placed adjacent to the item (not on the item, except on cable assemblies).

Definition of *Functional Designation*: Words, abbreviations, or meaningful number or letter combinations, usually derived from the function of an item (for example, slew, yaw), used on drawings, instructional material, and equipment to

© Springer International Publishing Switzerland 2015
R. Hanifan, *Perfecting Engineering and Technical Drawing*, SpringerBriefs in
Applied Sciences and Technology 139, DOI 10.1007/978-3-319-06983-8_6

identify an item in terms of its function (A functional designation is not a reference designation nor a substitute for it).

In this chapter, I provide only basic reference designation information. For more advanced or specific applications, see the appropriate governing specifications, ASME Y14.44 and IEEE Std 315. If these standards are not used and you have created your own method of establishing reference designations then in all probability chaos will exist for users of the equipment. Also, developing a consistency in the assignment of reference designations is beneficial. There is a worldwide recognition of electrical items by their reference designation. Occasionally different reference designations are assigned for the same type of items. One person may use an "L" for a class letter of a terminal lug and others may use an "E". Always consider what the electrical function is. It is an electrical termination (E) not a lug! The ASME and IEEE standards are universal and have stood for many years. The letters identifying the class of an item shall be selected from IEEE Std 315. If you don't comply with these standards then this can be a violation of the contract and most importantly when the universal standard is not followed then interpretation by other users will be hindered and misinterpreted.

How Assigned

The appropriate reference designation is assigned to items *via the Parts List* (BOM) except for the top (end) item, which has no higher assembly or parts list. All electronic parts and equipment must have a reference designation, which shall appear on the parts list. Reference designations must agree on the assembly drawings and the appropriate electronic diagrams/wire lists.

Parts List

Reference designations are assigned by, and appear on, the parts list for all electronic parts of that particular assembly, and they ***do not carry any prefix***. For example, R1 (not A1R1), W1 (not A1W1), A1 (not A1A1). The numbers of the reference designation should be sequential. For example, A1, A2, A3, not A2, A4, A6.

Item Identifiers (Find Numbers) and Reference Designations

In the parts list, electronic items are assigned an item identifier (a "find number," and it normally is a number) and a reference designation. In some systems, the items are listed in the parts list by reference designation in lieu of an item identifier (find number). It is preferred to identify the location of the item in the field of the drawing by the reference designation and/or the find number (balloon) or a combination of both. Use of a reference designation only is especially beneficial in the case of circuit card assemblies, while electronic assemblies and cable assemblies may use a combination. Identification in the field of the drawing may be either in a wiring diagram (when one is included as part of the assembly) or in the item delineation. Using only the reference designation of bulk type items (such as splices, terminals, shield terminations, and such) that are identified in the wiring diagram with a reference designation is very beneficial and much more understandable than placing an item identifier (balloon) in the item delineation.

The only intent of an item identifier (find number) is to identify the location of an item in the assembly. The intent of a reference designation is the same, except it also identifies what type of an item it is and its location in a system in addition to its location at the level of installation.

Which Items Receive a Reference Designation?

Each electronic item/electronic assembly is assigned a reference designation. A reference designation defines the type of item it is and, by use of the *full* reference designation, it will also define exactly where it fits into the equipment.

For example, a cable is assigned W, resistors R, capacitors C, repairable electronic assemblies A, non-repairable electronic assemblies U, and so on. IEEE Std 315 defines the various reference designations approved for use (Table 6.1). This system has been in place since the beginning of time, or maybe 5 years before that, and it has never changed. This system makes items easily identifiable. It defines what an item is and where it is located in the equipment. A full reference designation is like an address on a map, and it will identify its exact location. Reference designation information is also used to create the mates with information on band markers of cable assemblies, and it is used in interconnect diagrams, test requirements, test specifications, schematics, family trees, and extensively in technical manuals and technical orders.

If you have no contract that invokes IEEE Std 315 or ASME Y14.44, then all you have are your own engineering procedures and/or common sense. See the listing of the most common reference designations. For a complete listing and definitions, see IEEE Std 315. Never create your own reference designation if one already exists, as this only creates confusion. These designations are universally understood and recognized.

Table 6.1 Commonly used reference designations (See IEEE Std 315 for complete listing)

Component	Reference Designation
Amplifier	AR
*Assembly (repairable)[a]	A
*Assembly (non-repairable)[b]	U
Cable assembly	W
Capacitor	C
Connector (receptacle connector)	J
Connector (plug connector)	P
Diode	CR or D
Fan (blower)	B
Filter	FL
Fuse	F
Heater	HR
Inductor	L
Jumper wire	W
Microcircuit	U
Motor	M
Oscillator	Y or G
Resistor, thermal	RT
Splice	E
Switch	S
Terminal	E
Terminal board/strip	TB
Test point	TP
Thermistor	RT
Transformer	T
Waveguide	W
Wire	W

[a]See IEEE Std 315, Para. 22.2.4 for definition of repairable when in doubt

[b]See IEEE Std 315, Para. 22.2.5 for definition of non-repairable when in doubt

Reference Designation Starts

Wherever a reference designation starts, it continues upward in the system, *with no omissions*. If there are any omissions then chaos is created in the reference designation system. When assigned erroneously an inappropriate reference designation can impact part marking, technical manuals, instruction manuals and assembly drawings.

A circuit card assembly (CCA) assigns reference designations to the various components on it. The assembly above that picks up the CCA and assigns a reference designation to it. The next assembly that picks up that assembly assigns a reference designation to that assembly, and so forth, until reaching the very top level. Do not assign a reference designation for an item at the item level. *The reference designation is assigned on the parts list at the level at which the item is installed.* For example. A drawing of a capacitor does not assign a reference designation at the detail level. The capacitor may have multiple usages, and the level at which it is installed will determine each reference designation.

Units, Sets, and Nomenclature

Top deliverable end items shall be assigned a unit number or a set number. It depends on the contract (Government), but usually it is a unit number. The contract (if a Government contract) defines whether it is a unit or set. Normally, there is a Unit 1, but there could also be a Unit 2, Unit 3, and so on (see Fig. 6.1). In the example of Fig. 6.1 if the item was to be comprised of more than one Unit then the complete reference designations would be prefixed with the appropriate unit number. At that point, all the assemblies will tree downward to the lowest level of electrical items. If it is a set, such as a navigation pod and a targeting pod, and if they are delivered as sets, then it would be Set 1 and Set 2 instead of units. The contract should define if it is to be Sets or Units. In addition, *assigned nomenclature* (not to be confused with drawing title) should be defined, or nomenclature may have to be requested. (See Mil-Std-196 or equivalent.)

The following are a few of the useful terms pertaining to units, sets, systems, etc.

Set A unit or units and necessary assemblies, subassemblies and parts connected together or used in association to perform an operational function. (e.g., radio receiving set; sound measuring set, which include parts, assemblies and units such as cables, microphone and measuring instruments, radar homing set.) ("Set" is also used to denote a collection of like parts such as a tool set, or a set of tires.)

Subsystem A combination of sets, groups, etc., which performs an operational function within a system and is a major subdivision of the system. An example is an Intercept-Aerial Guided Missile Subsystem.

System A combination of two or more sets, which may be physically separated when in operation, and such other assemblies, subassemblies and parts necessary to perform an operational function or functions. Some examples include an Integrated Shipboard Computer System and a Navigational Control System.

Nomenclature The combination of an item name and a type designator. These are defined as follows:

Item Name A name published in the Federal Cataloging Handbook H6, or that name developed by the requestor in accordance with ASME Y14.100, the portion applicable to drawing titles. Item names used with type designator assignments will be consistent with the policies of the Federal Cataloging Program. Examples of unacceptable item names include abbreviations, acronyms, descriptions of size, frequencies, etc.

Type Designator A combination of letters and numbers arranged in a specific sequence to provide a short significant method of identification.

Unit An assembly or any combination of parts, subassemblies, and assemblies mounted together normally capable of independent operation in a variety of situations (e.g., hydraulic jack, electric motor, electronic power supply, internal combustion engine, electric generator, radio receiver).

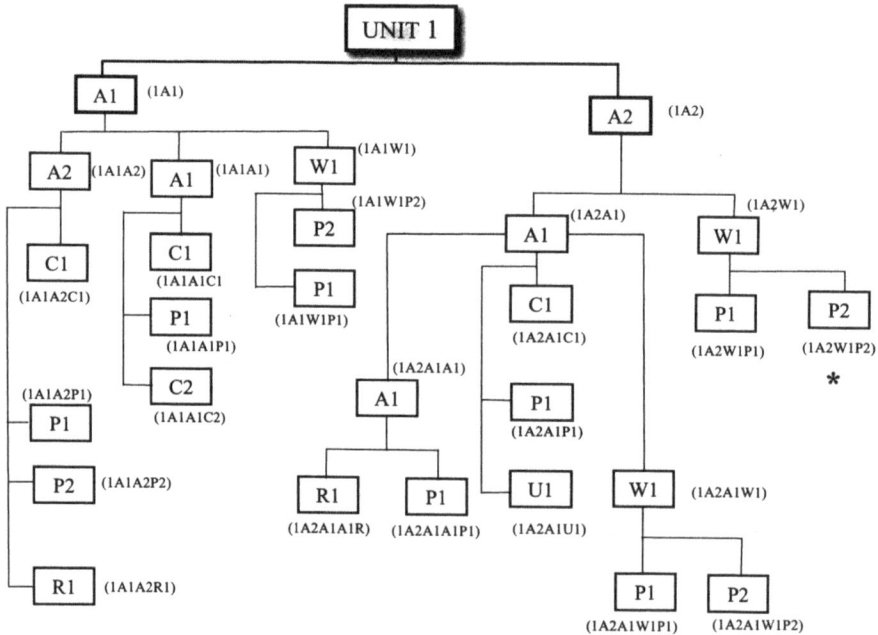

Fig. 6.1 Reference Designation Tree

Assigning Reference Designation

Start with the top assembly containing electrical parts and then assign a reference designation to each electronic item that appears on each assembly beneath it. Tree it all the way down until there are no more assemblies with electronic items (see Fig. 6.2).

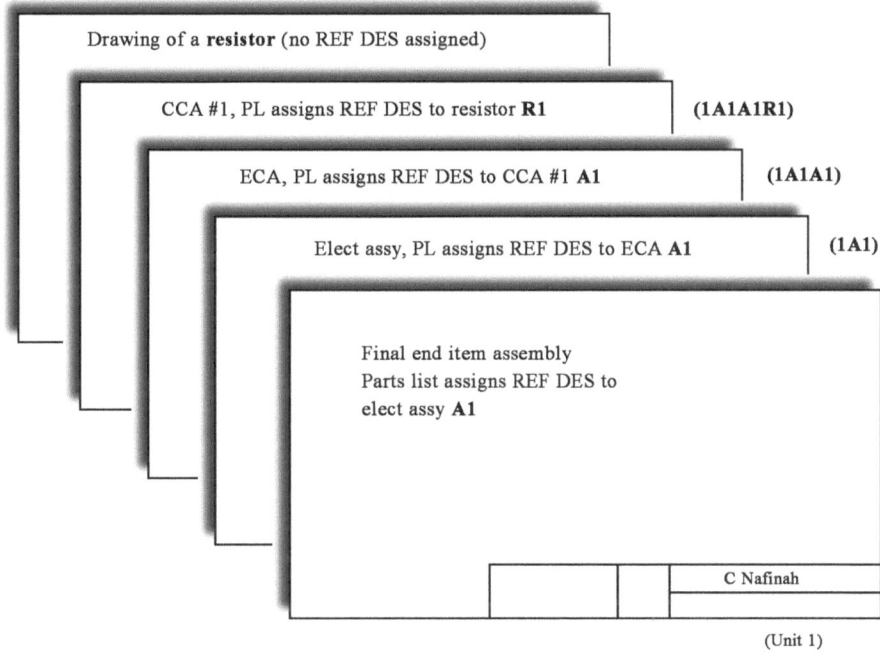

Fig. 6.2 Simplified version of flow of reference designations

Cable Assemblies

On cable assemblies, identify the "mates with" information. It is recommended to identify only as much of the prefix as is necessary to prevent improper hookups. Instead of "mates 1A1A1W1P1," "W1P1" is probably sufficient. Attempting to synchronize a connector reference designation to the reference designation of the mating connector is futile, and it really simplifies nothing. For example, assignment of a reference designation to a connector such as "P14" because it connects to "J14" is not beneficial. It just raises doubts of what happened to P1 through P13. With adequate "mates with" marking, the problem is easily resolved. Connector orientation should be shown when required.

Cable assemblies normally contain a "Connection Diagram" (Wiring Diagram) or a "Wire List", but both should not be prepared. A connection diagram is preferred provided that the information for a "Connection Diagram" is provided and clearly defined. See "Connection Diagram" as follows for the required information.

Connection Diagrams or Wiring Diagram

Connection diagrams depict the general physical arrangement of electrical connections and wires between circuit elements in an installation or assembly. It shows internal connections, but may include external connections that have one termination inside and one outside the assembly. It contains the details necessary to make or trace connections involved.

Reference designations.
Physical relationship of circuit elements and their connections.
Terminal arrangements.
Wires numbered for reference.
Wire and termination descriptions.

Reference ASME Y14.24 and ASME Y14.44 for complete drawing requirements for Cable Assemblies, Connection Diagrams, Wire Lists and Reference Designations.

Marking

Marking of reference designation designations is normally accomplished by placing the marking adjacent to the item and not directly on the item. ASME Y14.44 establishes the formulation and identification requirements of reference designations for electrical and electronic parts and equipment.

Connector, Wiring Side

Usage of the reference designation "X" denotes wiring side. The prefix "X" is normally used to denote the wiring side or the side where an item plugs in. It is not a "*stand-alone*" reference designation but only the indicator used to identify a location where the item plugs in. For example, connector P1 plugs into XP1. The connector is still referred to as P1, not XP1. XP1 identifies the location where P1 is installed.

Partial Reference Designations

Reference designations, assigned to the parts list and shown on the drawing, are only partial reference designations, as they require all the reference designations above them to be a complete and full reference designation. To clarify this, *it is required* and common practice to place a note on the drawing such as:

> Partial reference designations are shown. For full reference designation, prefix the reference designations with the unit number (or set number) and subassembly designation.

Chapter 7
Composite Laminate Drawings

There are no published design or drawing requirements for Composite Laminate Drawings and the rules are normally established by the design activity. The following are a few recommendations of procedures that I have found useful.

Composite drawings are normally prepared as Inseparable Assembly Drawings. Commonly they are of complex surfaces that are difficult to describe dimensionally and are frequently prepared as Reduced Dimensions Drawings (RDD) which invoke the model to be used to ascertain the dimensions. This drawing specifies sufficient datums configuration dimensions and tolerances to define the end item. Include sufficient dimensions and tolerances to define the location of plies only when these features affect the strength or final product of the finished part. Include sufficient dimensions and tolerances to define the ends of plies and overlaps of plies (or where these features may not be located) only when these features affect the strength or final product of the finished part. See Figs. 7.1 and 7.2.

Include a reference to the Reduced Dimension Drawing (RDD), including appropriate tolerances (Usually a profile), when required to define the outside contour (mold lines) and taper regions which would be impractical to dimension on a drawing.

Use find numbers in the drawing field to locate parts defined on other drawings that are used in the composite laminate and for details other than ply parts. Specify ply material with find numbers on the drawing and Parts List. Quantify the specific materials in the Parts List with an "AR". Do not identify the separate plies with either a Find Number or Part Identification Number and do not illustrate the flat pattern design for any of the plies.

The drawing should only disclose enough data and requirements for a manufacturing function to be able to create interchangeable hardware and document the specific method selected to fabricate parts.

Include sufficient dimensions and tolerances to define the location of each ply and ply part.

© Springer International Publishing Switzerland 2015

R. Hanifan, *Perfecting Engineering and Technical Drawing*, SpringerBriefs in Applied Sciences and Technology 139, DOI 10.1007/978-3-319-06983-8_7

It is recommended to include a note such as "UNLESS OTHERWISE SPECI-
FIED, LOCATING TOLERANCE FOR ALL PLIES IS .XX." Exceptions should
be indicated by specific note or drawing dimension and tolerance. Title block
tolerances should apply to flat pattern dash numbers, unless otherwise noted.

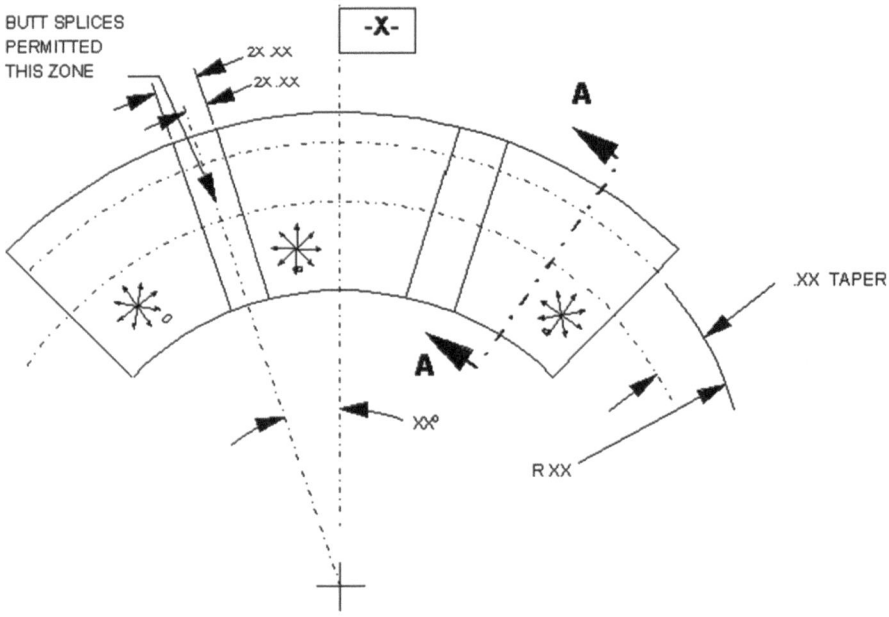

Fig. 7.1 Example of sufficient dimensioning of plies

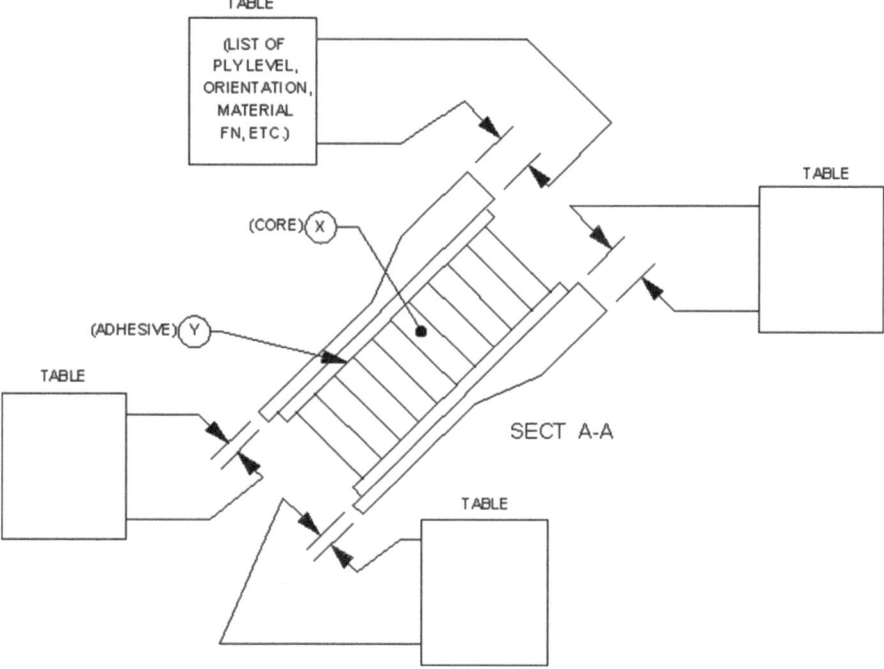

Fig. 7.2 Example of sufficient dimensioning of ply description and orientation

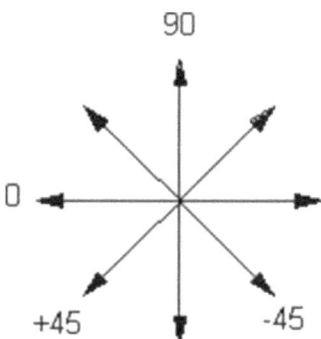

Fig. 7.3 Orientation rosette

Show this orientation rosette on the main view and on auxiliary views as necessary to adequately define fiber orientation of each ply part by the use of a rosette to that shown in Fig. 7.3. Indicate in the general notes that the fiber orientations are in degrees.

Orientation Rosette

Identify the arrangement of the plies by ply number on the end views and/or sections.

(a) The plies must be legible and discernible. When necessary, the thickness of the ply may be exaggerated.
(b) Assign ply levels in numerical sequence, e.g., 1, 2, 3,, 4, etc., starting from one face of the laminate (usually the mold line face or the tool side).
(c) Specify the position of the plies in each section by use of a ply table as shown in Fig. 7.4. Include the ply level and the (ply part's flat pattern dash number when flat pattern drawings are made) and fiber orientation. Indicate the start and end plies by leaders and arrows from the appropriate table entries.

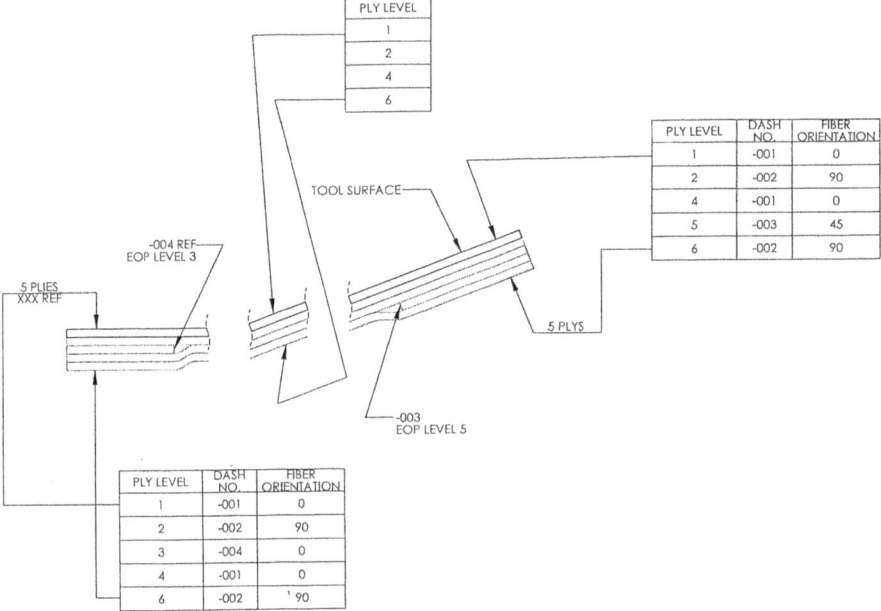

Fig. 7.4 Ply tables

When a ply must be added between existing plies, add a single alpha suffix to the preceding (lower) ply level and each intervening ply. For example, if two plies are to be added between ply level 170 and ply level 171, ply level 170 becomes level 170A and the added plies are identified as 170B and 170C.

Indicate the end of a ply (EOP) in the section by a vertical leader or a leader normal to the ply being terminated so as to avoid ambiguous interpretations.

NOTE: The acronym "EOP" (End Of Ply) must be defined in the general notes.

Include a table on the first sheet that specifies which sheet of the drawing depicts which ply flat pattern.

Whenever possible, align the ply views in the same direction as the ply position in the assembly views to provide the proper relationship of the fiber orientation. Fiber orientation must be defined relative to the lay-up of the finished item.

Identify each ply part with its flat pattern dash number. Locate the dash number within the outline of the part except a leader may be used when the part's size does not allow inclusion of the dash number within the outline.

Indicate the fiber orientation of each ply with a directional arrow and an angular expression next to the (flat pattern dash number [only when flat pattern drawings are made] enclosed as shown in Fig. 7.5. When a particular orientation is not required, state "OPTIONAL".

Fig. 7.5 Identification of ply laminate parts

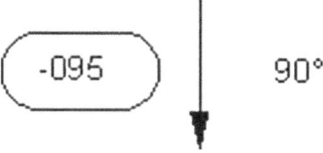

Bi-directional material (i.e., material with equal or near equal strength in both directions, [warp and fill]) fiber orientation may be expressed "0/90 DEGREES" or "+45 DEGREES" as applicable.

The strongest fibers (normally warp) establish the fiber orientation when using bi-directional material not equal in strength in both directions.

A ply part may be used in more than one ply level or more than once in a ply level but must always be depicted exactly the same size. Parts with the same configuration, but with different fiber orientation will have different detail dash numbers.

When splices are permissible, always state that fact in the drawing notes.

When splices are permitted, but the splice location is mandatory, add a note controlling the permissible location and specifying the requirement to stagger ply butts at the splice.

The following is a listing of the most commonly used words and their definitions used on composite drawings.

Definitions

Bi-Directional Woven Material	A material with a structural fiber in both warp and fill direction.
Composite Material	Composite materials are combinations of materials differing in composition or form. The constituents retain their identities in the composite; that is, they do not

dissolve or otherwise merge completely into each other although they act in concert. Normally, the components can be physically identified and exhibit an interface between components.

Fill	Fibers transverse to the warp.
Fiber	A single homogenous strand of material with high axial strength used as a principle constituent in composites.
Fiber Direction/ Orientation	Direction of filament axis (or fabric warp) in the ply part. This is generally established as an angular relationship to the principle load direction. Bi-directional woven material with equal or nearly equal strength in both directions may have a dual fiber orientation, e.g., 0 degrees/90 degrees, plus or minus 45 degrees.
Flat Pattern Dash Number	A suffix number to the drawing number that, together with the drawing number, identifies a discrete piece of material (ply part) forming a ply of a laminate or one of several distinct pieces forming a ply of a laminate. The dash number through its CAE representation and Parts List data defines the size, shape, material, and fiber orientation of the ply part.
Lamina	A single ply or layer in a laminate.
Laminae	The plural of lamina.
Laminate	A product made by bonding two or more laminae of material or materials.
Lay-Up	A process of fabrication involving the placement of Successive laminae.
Matrix	The essentially homogeneous material in which the fibers of filaments of a composite are imbedded.
Ply	A single lamina comprised of one or more discrete pieces (ply parts) of fabrics (e.g., fiber glass cloth) and/or filaments (e.g., boron or graphite).
Ply Level	An integer number assigned to a lamina, counted from a reference surface that is usually the tool surface.
Ply Number	The numerical identification given to each ply.
Unidirectional Material	A material in which all fibers lie in a single direction.
Warp	The lengthwise fibers in a loom.

Chapter 8
Inspection

The Integrity of Your Supplier: Who Inspects Parts and What Is Inspected

Inspection is always a gamble. Which features are to be inspected and which parts shall be inspected. There really is no requirement to inspect parts or established inspection rules. Inspection is derived by a confidence in your fabricator or vendor. The reality is that most parts will receive only cursory inspections by the procuring activity (design activity). Most inspections will be performed by the fabricator or vendor. The reasons for this vary greatly, but it is primarily dependent on that availability of equipment for inspectors in the field. Many features may go uninspected and 100 % inspection is also not realistic. Many of the uninspected features are noncritical, but critical features frequently may be over-looked. Incoming (in house) inspection at the procuring activity has more capability, but in many cases this is just too late, and inspection/rejection causes more delays. In all probability, parts will most likely be reworked in house instead of rejecting them back to the fabricator or vendor. When parts are rejected at the procuring activity, many delays, or even line stoppages can be incurred.

A drawing normally does not specify any inspection criteria. What is inspected, if anything is entirely up to the procuring activity and the confidence you have in your fabricator or vendor. You may elect to have full inspection until you have developed sufficient faith in the fabricator/vendor. Any inspection less than 100 % is strictly a gamble, and you have to feel confident for whatever level of inspection you perform.

It is recommended that *critical features be identified* in some manner and denoted on the drawing or in a separate document (an inspection plan or the like). In this document, denote which specified features shall receive a 100 % inspection,

© Springer International Publishing Switzerland 2015
R. Hanifan, *Perfecting Engineering and Technical Drawing*, SpringerBriefs in
Applied Sciences and Technology 139, DOI 10.1007/978-3-319-06983-8_8

75 % inspection, 50 % inspection, and so forth. When this data is specified in a separate inspection plan, it can reduce the amount of change activity to the drawing. When the data is placed on the drawing, an engineering change will be required to make any revisions. It is much easier to change one inspection procedure than 100 drawings.

Chapter 9
Delivery of Your Drawings (Technical Data Package)

In this chapter, I will explain how delivery of the drawings that you prepared impacts the delivery of the technical data package to the customer. If you are not on a Government program that requires delivery, then this chapter is not significant to you. In past years, almost all Government contracts required delivery, and now this requirement has been reduced to minimal delivery requirements. There is always a movement to revert back to the days of the past invoking delivery requirements, but this usually occurs when the customer foresees reprocurement from other sources.

There are many variables in the delivery of the technical data package (TDP) and in the drawing preparation requirements. Always be aware of what your requirements are, either internal or customer requirements.

This is not the first chapter of the book; however, it should be treated as a "read me first" type of subject. Your standard drawing practices and the following topics *should have been discussed at the beginning of the contract or at the time of the contract award to avoid problems at the time of drawing delivery*. Clarifying your practices with the customer at the beginning can minimize the most common discrepancies and misunderstandings that may occur at the time of delivery of the technical data package. If these items are resolved or clarified in the beginning with the customer, huge savings may result, and transmittal of the technical data package will be facilitated. Careful scrutiny of the contractual requirements and your ability to comply with these requirements at the beginning of the contract allows you to plan accordingly in your documentation and subcontracting efforts. Never be in the situation at the end of a program and realizing that your practices or subcontractors are in violation of the contract.

The following are the most significant problems that require resolution or close attention during the preparation of the engineering documentation so as to minimize their impact at the time of delivery.

© Springer International Publishing Switzerland 2015

R. Hanifan, *Perfecting Engineering and Technical Drawing*, SpringerBriefs in Applied Sciences and Technology 139, DOI 10.1007/978-3-319-06983-8_9

Vendor Prints

If you used vendor part numbers anywhere in your documentation, you may be expected to provide a copy of those drawings, or is your customer expected to obtain them? Will they comply with the legibility and documentation requirements desired/required by the customer? Are these documents expected to comply with the same level of drawing requirements as specified in the contract, or will they be acceptable in the as received status? If you know the answer to these questions at the time of the contract award, much grief and expense can be avoided. Attempting to rectify erroneous interpretations of what the customer desires at the end of the program may result in huge manpower and money expenditures.

Missing Drawings

This always sounds like an insignificant item, and it probably is nothing more than a logistical problem. However, in some cases it may be a deeper rooted problem than it appears. The customer will require a copy of every drawing or document that is called out directly and on sub-tier documents. If it is called out on the drawing or sub-tier documents, the customer will probably demand a copy of the document! Confusion between specifications and drawings always abounds, which normally is a result of documentation practices.

Design activities always want to make word drawings (which they commonly refer to as specifications) look like specifications, because it looks professional. They process these specifications like drawings and placed them on drawing formats. True specifications are prepared per Mil-Std-961 and are normally delivered as separate elements of the contract, and as specified in the Contract Data Requirements List (CDRL). *If it is on a drawing format, it is **not a specification***, it is a drawing! If it is looks like a *book and is not on a drawing format, then it is probably a specification*. This delivery of specifications is normally by either hardbound paper delivery or in the form of digital data. The problems begin when these specifications are listed on drawings or appear on data lists under the heading of drawings. This confuses the customer, and then he expects to see the specifications delivered along with the rest of the drawing technical data package and in the same delivered form.

Undisclosed Data

Undisclosed data is another area to be avoided. The reference to unique, undisclosed information (such as test fixtures, test circuits, digital data, procedures, and so forth) does not fulfill the requirements of the contract and when discovered at the termination of the contract, may result in delayed payment by the customer until this information is provided.

Delivery of Digital Data and Models

Ensure that digital data and models have been prepared and the delivery is done per the contract. Also, ascertain whether the customer can receive, store, and retrieve this data as necessary.

Delivery of the technical data package is usually the final phase of your engineering efforts, and that is normally when the program gets paid the remainder of the contract money. Additional deliveries will probably be required to provide the balance of drawings, or as changes occur to the data package on a prearranged scheduled basis, such as every 90 days.

Many problems can occur, but careful scrutiny of missing drawings, undisclosed data, and vendor prints can eliminate the majority of the problems.

Chapter 10
The Checker: What He Will Look for

The checker that reviews your drawing has a different mindset from yours. You are optimistic, your drawing is right and there is not a mistake on it, it is perfect. However, the checker knows your drawing is not perfect but that there is something right on it and he just has to find it. He will sift through your drawing notes, the delineation, geometric dimensioning and tolerancing, and the parts list. All data is reviewed for correctness and contractual compliance. He has to evaluate any errors and their significance. If petty and relatively unimportant, he may discard them. He has a mental checklist that he uses, and it is different for every drawing type. A competent checker can find 75 % of your errors in less than a half an hour, and if you know what he looks for, then you will be many steps ahead.

Checking is always a very difficult task and takes a very special individual. He must overcome his desire to redesign as well as being flexible for the various types of checking needs. He may be asked to perform a full-up check, design analysis, tolerance stack up, or verify that all contractual requirements are met for a production level contract. He may be expected to perform only a quick check for specific requirements for a very limited production, or he may be looking at just a quick drawing created to obtain quotes. He must be willing to delve into intricate details until he is completely satisfied that all errors have been discovered.

The more cooperative the designer is in providing information, the better the end item will be. The better the end item, then the better you will look. No one may remember that you did a good job, but they will definitely remember your name if you did a bad job.

It would be beneficial if you would sit with the checker prior to preparing the drawing and discuss it with him. Presenting all your backup data and discussing your datum structure and dimensioning scheme with him will benefit you both. You may want to discuss the type of drawing selected, materials and finishes used and information of the mating parts and their sizes. This simple act will facilitate the processing of your drawing and resolve many errors and misinterpretations before they become a problem. The more that the checker/reviewer understands of the design the more accurate his check will be.

© Springer International Publishing Switzerland 2015
R. Hanifan, *Perfecting Engineering and Technical Drawing*, SpringerBriefs in Applied Sciences and Technology 139, DOI 10.1007/978-3-319-06983-8_10

The first item that a checker will need to determine is what the proper drawing requirements are. Is it for a contract, or is it to in house requirements? He will then start a systematic check. His mental checklist may be as follows (it will vary for each particular type of drawing, and this is only a partial listing):

- Check drawing title, sheet numbering, revision information, title block information, including contract number (if applicable), appropriate notations ("see separate parts list," etc.) and drawing statements (limited rights, export notices, etc.).
- Parts list is complete and agrees with the drawing. Reference designations have been assigned to electrical items.
- Drawing and model versions agree with the CAD system.
- CAD models will be investigated to uncover problems to the level requested. It may be model and layer naming conventions, status, interferences, and compliance with all company procedures.
- Verify that all features are dimensioned and have tolerances.
- Notes are complete and unambiguous, and all information contains acceptance and rejection criteria.
- Material and finishes are completely specified, including such information as types, classes, temper, and appropriateness for the environments of design.
- Datums selected and shown appropriately. Datums are logical, have sufficient size, and are contactable.
- Features are located properly with geometric dimensioning and tolerancing, and it has been applied properly:

  - Features located to or from a centerline have been properly defined and dimensioned.
  - Modifier is applied properly, and MMC has not been used just as a default but considered appropriately in the tolerance stack up.
  - Tolerancing is applied appropriately.
  - Data in the feature control frame is properly presented.
  - Hole size and tolerance are appropriate for standard drill sizes. Hole tolerance has been calculated properly or has positional tolerance and .014 been used just as a default.
  - Surfaces having all elements in the same plane (coplanar) have been appropriately toleranced.
  - Features that require angular orientation have been defined properly.
  - If painted, is it denoted that dimensions and tolerances apply after plating and prior to painting.
  - Corner and fillet radius are specified and shown properly.
  - Deburr and sharp edges note, does it conflict with corner and fillet radii?
  - Tolerances are not overly stringent and are appropriate for design. Angular dimensions—is the tolerance too restrictive?
  - Mating parts and interfaces have been verified, and no interferences are present.
  - There are no other design flaws (interferences) present.

- Control drawings have no unnecessary dimensional restrictions.
- Assembly drawings have all their parts shown with appropriate reference designations assigned for electrical items.

Chapter 11
Other Information

Definition of Common Terms

Understand some of the basic terms so that, when specified, they are used appropriately. If used incorrectly, they will cause confusion.

An example that is frequently seen is, "Inspection to be performed by contractor."

Should it have stated? "Inspection to be performed by procuring activity."

Contractor	A company having a contract with the procuring activity for the design, development, manufacture, maintenance, modification, or supply of items under the terms of the contract.
Datum	A theoretically exact point, axis, or plane derived from its geometric counterpart of a specified datum feature. A datum is the origin from which the location or geometric characteristic of features of a part are established. In simpler terms, a feature on the actual part (datum feature) is used to create the actual datum, and it is the origin from which the dimensions will be inspected. The actual feature is simulated in the processing equipment. For example, it may be the surface plate or machine table that the item is sitting on, or if the datum feature is a hole, then the actual datum may be the pin. Datums should always be selected by functionality and accessibility.
Design Activity	An activity having the responsibility for design of an item.
Original Activity	The activity that had original responsibility for the item.
Current Design Activity	The activity currently having the responsibility for design and maintenance.

© Springer International Publishing Switzerland 2015
R. Hanifan, *Perfecting Engineering and Technical Drawing*, SpringerBriefs in
Applied Sciences and Technology 139, DOI 10.1007/978-3-319-06983-8_11

Least Material Condition	The condition in which a feature of size contains the least amount of material within the stated limits of size. For example, maximum hole or minimum shaft diameter.
Maximum Material Condition	The condition in which a feature of size contains the maximum amount of material within the stated limits of size. For example, minimum hole diameter or maximum shaft diameter.
Part	One item, or two or more items joined together, that is not normally subject to disassembly without destruction or impairment of the designed use.
Procuring Activity	The contracting activity or procuring activity is the customer.
Regardless of Feature Size	The term used to indicate that a geometric tolerance or datum reference applies at any increment of size of the feature within its size tolerance.
Unit	An assembly, or any combination of parts, subassemblies, and assemblies mounted together, normally capable of independent operation in a variety of situations.
Vendor	A source from whom a purchased item is obtained (Tables 11.1, 11.2, 11.3 and 11.4).

Table 11.1 Standard drill size

Drill Size	Dec. Equiv.	Drill Size	Dec. Equiv.	Drill Size	Dec. Equiv.	Drill Size	Dec. Equiv.
80	.0135	43	.089	8	.199	X	.397
79	.0145	42	.0935	7	.201	Y	.404
1/64	**.0156**	**3/32**	**.0938**	**13/64**	**.2031**	13/32	.4062
78	.016	41	.096	6	.204	Z	.413
77	.018	40	.098	5	.2055	27/64	.4219
76	.020	39	.0995	4	.209	**7/16**	**.4375**
75	.021	38	.1015	3	.213	29/64	.4531
74	.0225	37	.104	2	.221	**15/32**	**.4688**
73	.024	36	.1065	1	.228	31/64	.4844
72	.025	**7/64**	**.1094**	A	.234	**1/2**	**.500**
71	.026	35	.110	**15/64**	**.2344**	33/64	.5156
70	.028	34	.111	B	.238	**17/32**	**.5312**
69	.0292	33	.113	C	.242	35/64	.5469
68	.031	32	.116	D	.246	**9/16**	**.5625**
1/32	**.0312**	31	.120	**1/4**	**.250**	37/64	.5781
67	.032	**1/8**	**.125**	F	.257	**19/32**	**.5938**
66	.033	30	.1285	G	.261	39/64	.6094
65	.035	29	.136	**17/64**	**.2656**	**5/8**	**.625**
64	.036	28	.1405	H	.266	41/64	.6406
63	.037	**9/64**	**.1406**	I	.272	**21/32**	**.6562**
62	.038	27	.144	J	.277	43/64	.6719
61	.039	26	.147	K	.281	**11/16**	**.6875**
60	.040	25	.1495	**9/32**	**.2812**	45/64	.7031
59	.041	24	.152	L	.290	**23/32**	**.7188**
58	.042	23	.154	M	.295	47/64	.7344
57	.043	**5/32**	**.1562**	19/64	.2969	**3/4**	**.750**
56	.0465	22	.157	N	.302	49/64	.7656
3/64	**.0469**	21	.159	**5/16**	**.3125**	**25/32**	**.7812**
55	.052	20	.161	O	.316	51/64	.7969
54	.055	19	.166	P	.323	**13/16**	**.8125**
53	.0595	18	.1695	21/64	.3281	53/64	.8281
1/16	**.0625**	**11/64**	**.1719**	Q	.332	**27/32**	**.8438**
52	.0635	17	.173	R	.339	55/64	.8594
51	.067	16	.177	**11/32**	**.3438**	**7/8**	**.875**
50	.070	15	.180	S	.348	57/64	.8906
49	.073	14	.182	T	.358	**29/32**	**.9062**
48	.076	13	.185	23/64	.3594	59/64	.9219
5/64	**.0781**	**3/16**	**.1875**	U	.368	**15/16**	**.9375**
47	.0785	12	.189	**3/8**	**.375**	61/64	.9531
46	.081	11	.191	V	.377	**31/32**	**.9688**
45	.082	10	.1935	W	.386	63/64	.9844
44	.086	9	.196	25/64	.3906		

Table 11.2 Standard drill hole tolerances from AND 10387 (Inactive)

Hole Dia.	Tolerance
.0135–.125	+.004/–.001
.126–.250	+.005/–.001
.251–.500	+.006/–.001
.501–.750	+.008/–.001
.751–1.000	+.010/–.001
1.001–2.000	+.012/–.001

Table 11.3 Stock thicknesses

Aluminum	Steel	Aluminum	Steel
.010	.0149 (28 GA)	.125	.1196 (11 GA)
.012	.0179 (26 GA)	.160	.1345 (10 GA)
.016	.0239 (24 GA)	.190	.1644 (8 GA)
.020	.0299 (22 GA)	.250	.1793 (7 GA)
.025	.0359 (20 GA)	.312	.25
.032	.0418 (19 GA)	.375	.312
.040	.0478 (18 GA)	.50	.375
.050	.0598 (16 GA)	.75	.50
.063	.0747 (14 GA)	.875	.75
.080	.0897 (13 GA)	1.00	.875
.090	.1046 (12 GA)		1.00

Table 11.4 AWG sizes

American wire gage size			
AWG size	Conductor DIA	AWG	Conductor DIA
000	.409	17	.045
00	.365	18	.040
0	.325	19	.035
1	.289	20	.032
2	.257	21	.028
3	.229	22	.025
4	.204	23	.022
5	.181	24	.020
6	.162	25	.017
7	.144	26	.015
8	.128	27	.014
9	.114	28	.012
10	.101	29	.011
11	.090	30	.010
12	.080	31	.0089
13	.072	32	.008
14	.064		
15	.057		
16	.050		

To Calculate a Positional Tolerance

See Fig. 11.1.

Take maximum material condition (MMC) of hole (smallest hole)

Subtract maximum material condition (MMC) of fastener (largest fastener diameter)

Remainder is positional tolerance available.

If it is a floating fastener then that is what is available for each part.

If it is a fixed fastener then that is the *total tolerance* available for each part.

EXAMPLE: FLOATING FASTENER

Hole size .281 +.006 -.001   MMC of hole is   .280

Fastener   .250 +.000 -.001   MMC of fastener is .250

TOLERANCE AVAILABLE   .030

If it is a floating fastener then .030 tolerance is available for each part.

If it is a fixed fastener then .030 ***total*** tolerance is available (or .015 tolerance each part. The .030 tolerance may be divided equally or non-equally between the two parts. Example one part may have .020 tolerance and the other .010 tolerance.)

Fig. 11.1 Positional tolerance calculation

		THREADS PER INCH								
		SERIES WITH GRADED PITCHES			SERIES WITH CONSTANT PITCHES					
SIZE	MAJ DIA	COARSE UNC	FINE UNF	EXRA FINE UNEF	8UN	12 UN	16 UN	20 UN	28 UN	32UN
0	.060		80							
1	073	64	72							
2	.086	56	64							
3	.099	48	56							
4	.112	40	48							
5	.125	40	44							
6	.138	32	40							UNC
8	.164	32	36							UNC
10	.190	24	32							UNF
12	.216	24	28	32					UNF	UNEF
¼	.250	20	28	32				UNC	UNF	UNEF
5/16	.3125	18	24	32				20	28	UNEF
3/8	.375	16	24	32			UNC	20	28	UNEF
7/16	.4375	14	20	28			16	UNF	UNEF	32
½	.500	13	20	28			16	UNF	UNEF	32
9/16	.5265	12	18	24		UNC	16	20	28	32
11/16	.6875			24		12	16	20	28	32
¾	.750	10	16	20		12	UNF	UNEF	28	32
7/8	.875	9	14	20		12	16	UNEF	28	32
1	1.00	8	12	20	UNC	UNF	16	UNEF	28	32

Fig. 11.2 Thread sizes

Conclusion

By paying attention to the areas that have been shown to cause most drawing and design errors, these problems can be reduced to a more acceptable level. Hopefully, drawing, change activity, production, and inspection costs will be reduced.

With special attention to the following areas, your endeavors will be more rewarding and you will have reduced your errors. Some organizations use check lists, however I dislike checklists immensely, although others swear by them. I feel that when one will uses a checklist you are prone to only checking what is listed on the check list and not think outside of the box. Conditions differ on every drawing and no checklist can cover all situations and drawing types. I have seen companies that devoted more time to the proper preparation of a checklist than they did to the drawing, even though the checklist had minimal relevance to the drawing. A checklist can be beneficial if it is only used for simple items such as drawing title, number, used on information and so forth.

The following is not a checklist but just a reminder of the more important areas.

If you are dimensioning to or from a centerline, is it defined?

If you are dimensioning to or from a centerline, try to visualize it as an axis and not as a crosshair.

If applicable, is angular orientation of the feature required?

Are you using a maximum material modifier as a default without understanding the impact?

Does all data contain acceptance and/or rejection criteria? Check your notes especially.

Do you know what your design and documentation requirements are? Are you following your engineering procedures, or do you have a contractual requirement?

If you have a contractual requirement, do you know what is required?

© Springer International Publishing Switzerland 2015
R. Hanifan, *Perfecting Engineering and Technical Drawing*, SpringerBriefs in Applied Sciences and Technology 139, DOI 10.1007/978-3-319-06983-8

A Final Thought or Two

Always understand all the information that you placed on the drawing. If there is ever any information that you don't understand, don't be afraid to ask. So often I have seen people who don't want to ask or are afraid to ask because they don't want to look dumb, and when they finally get the courage to ask, they find out that others also don't know either. Just like you, the others were afraid to ask. If you have a question, why not ask the engineering design checker? He will love it, especially since he won't have to mark up a misapplication on your drawing.

Best of luck and Semper Fi.

© Springer International Publishing Switzerland 2015

R. Hanifan, *Perfecting Engineering and Technical Drawing*, SpringerBriefs in Applied Sciences and Technology 139, DOI 10.1007/978-3-319-06983-8